Ultra-Wideband Pulse-based Radio

ANALOG CIRCUITS AND SIGNAL PROCESSING SERIES

Consulting Editor: **Mohammed Ismail.** ***Ohio State University***

For other titles published in this series, go to
http://www.springer.com/series/7381

Wim Vereecken · Michiel Steyaert

Ultra-Wideband Pulse-based Radio

Reliable Communication over a Wideband Channel

Springer

Wim Vereecken
Katholieke Universiteit Leuven
Dept. Electrical Engineering
(ESAT)
Kasteelpark Arenberg 10
3001 Leuven
Belgium
wim.vereecken@esat.kuleuven.be

Michiel Steyaert
Katholieke Universiteit Leuven
Dept. Electrical Engineering
(ESAT)
Kasteelpark Arenberg 10
3001 Leuven
Belgium
michiel.steyaert@esat.kuleuven.be

ISBN 978-90-481-2449-7 e-ISBN 978-90-481-2450-3
DOI 10.1007/978-90-481-2450-3
Springer Dordrecht Heidelberg London New York

Library of Congress Control Number: 2009926324

Printed on acid-free paper

Springer is part of Springer Science+Business Media (www.springer.com)

Contents

Preface

Today's booming expanse of personal wireless radio communications is a rich source of new challenges for the designer of the underlying enabling technologies. Personal communication networks are designed from a fundamentally different perspective than broadcast service networks, such as radio and television. While the focus of the latter is on reliability and user comfort, the emphasis of personal communication devices is on throughput and mobility. However, because the wireless channel is a shared transmission medium with only very limited resources, a trade-off has to be made between mobility and the number of simultaneous users in a confined geographical area. According to Shannon's theorem on channel capacity,[1] the overall data throughput of a communication channel benefits from either a linear increase of the transmission bandwidth, or an (equivalent) exponential increase in signal quality. Consequently, it is more beneficial to think in terms of channel bandwidth than it is to pursue a high transmission power. All the above elements are embodied in the concept of *spatial efficiency*. By describing the throughput of a system in terms of bits/s/Hz/m^2, spatial efficiency takes into account that the use of a low transmission power reduces the operational range of a radio transmission, and as such enables a higher reuse rate of the same frequency spectrum.

What is not accounted for in the above high-level theoretical perspective, is that a wide transmission bandwidth opens up a Pandora's box of many complications at receiver side. Shannon's theorem is indeed valid for an AWGN channel, but the environment where network devices are operated in, usually refuses to fit this idealized model. A real-world channel, for example, will suffer from multipath reflections: multiple, delayed versions of the same trans-

[1]Channel capacity = bandwidth $\times \log_2(1 + \text{signal quality})$.

mission arrive at the receive antenna and start to interfere with one another, an effect that is known as *intersymbol interference*. Apart from this form of self-interference, a wide transmission band is also a wide open door for in-band interfering signals. It is not the presence of the interferer itself that causes the problem, but the sometimes very large difference in the power balance between the unwanted component and the signal-of-interest. By putting considerable stress on the linearity requirements of the receiver, high-powered interferers indirectly impact the battery lifetime of portable devices.

This work lays the foundations of a new radio architecture, based on the pulse-based radio principle. As will become clear throughout this book, using short pulses with a wide spectral footprint has considerable advantages for the reliability of a wireless transmission under indoor channel conditions. Notwithstanding being described as a pulse-based system, the presented architecture is also a direct descendant of single-carrier QPSK modulated radio. This genealogical line ensures the system can enjoy the best of both worlds: a high reliability and a fairly uncomplicated modulation technique.

However, simplicity does not preclude powerful capabilities. From the very early stages on, the high-level system design was conceived with the above described complications of the wideband radio channel in mind. Issues that come with the unpredictable nature of the wireless medium, such as interference and varying channel conditions, are dealt with at multiple levels in the system hierarchy. For example, a specially crafted *interferer suppression and signal reconstruction* algorithm has been developed (chapter 3). Without active intervention from the transmitter, the ISSR system – which is located entirely at receiver side – is capable of on-the-fly cleaning of frequency bands which have fallen victim to multipath fading or narrowband interference. The unique blend of pulse-based radio, a simple modulation scheme and a powerful signal reconstruction system in the back-end make the presented pulse-based radio system a viable and promising alternative for the high-end (but highly complex) modulation schemes such as the OFDM-system currently widely adopted by WLAN applications.

As a proof of concept, the theoretical underpinnings of this work are supported by the implementation of an analog front-end for pulse-based radio in 0.18 μm CMOS. The quadrature RF front-end comprises a wideband RF input stage, an I/Q pulse-to-baseband downconversion mixer and a variable gain amplifier (the latter based on a novel highly-linear open-loop topology). The prototype chip has drawn attention to some subtle technical issues inherent to pulse-based radio. For example, the sensitivity of the receiver may be adversely affected by leakage of clock signals into the sensitive signal chain. While this effect does not come to the surface in high-level system simulations, it can be easily prevented by some simple precautions in the early stages of the design process.

As a conclusion, the chip-level realization did not only prove the feasibility of a quadrature pulse-based transceiver system, but has also marked some key points that need special attention from a developer's viewpoint during the design of a pulse-based radio chipset.

Leuven, Belgium
October 2008

Wim Vereecken

List of Abbreviations and Symbols

Abbreviations

AC	Alternating current (commonly used in a small-signal context)
ADC	Analog-to-digital converter
AFC	Automatic frequency control
AGC	Automatic gain control
AM	Amplitude modulation
AM-SSB	Single sideband AM
AM-VSB	Amplitude modulation with a vestigial sideband (TV broadcast)
AWGN	Additive white Gaussian noise
BALUN	Balanced to unbalanced transformer
BER	Bit error rate
BICM	Bit interleaved coded modulation
BPF	Band-pass filter
BPSK	Binary phase shift keying
BW	Bandwidth
C/A	Coarse acquisition code (GPS related)
CCITT	Comité Consultatif International Téléphonique et Télégraphique
CCK	Complementary code keying
CDF	Cumulative distribution function
CDMA	Code division multiple access
CF	Crest factor
CL	Closed-loop
CMFB	Common-mode feedback

CML	Current mode logic
CMOS	Complementary metal oxide semiconductor
CMRR	Common mode rejection ratio
CNR	Carrier-to-noise ratio
CSMA/CA	Carrier sense multiple access/collision avoidance
CT	Confidence threshold
DAC	Digital-to-analog converter
DC	Direct current
DFT	Discrete Fourier transform
DIFS	Distributed interframe spacing (IEEE802.11 related)
DLL	Data link layer of the OSI model
DLL	Delay locked loop (PLL related)
DSP	Digital signal processing
DSSS	Direct sequence spread spectrum
E_b/N_0	Bit energy over noise density ratio
EGC	Equal gain combining
EIRP	Effective isotropic radiated power
ENOB	Effective number of bits
ERBW	Effective resolution bandwidth
ESD	Energy spectral density
ESD	Electrostatic discharge (protection circuit)
EVM	Error vector magnitude
FCC	Federal Communications Commission
FDMA	Frequency division multiple access
FEC	Forward error coding
FFT	Fast Fourier transform
FHSS	Frequency-hopping spread spectrum
FIR	Finite impulse response
FM	Frequency modulation
FSK	Frequency shift keying
FUND_x	Fundamental component at node x
GBW	Gain-bandwidth product
GMSK	Gaussian minimum shift keying
GPS	Global Positioning System
GSM	Global System for Mobile communications
$\text{HARM}_{n,x}$	n-th order harmonic component at node x
HD	Harmonic distortion
HD_2	Second-order harmonic distortion
HD_3	Third-order harmonic distortion
hop	Change of frequency in a FHSS transmission
IC	Integrated circuit
IDFT	Inverse discrete Fourier transform

IF	Intermediate frequency
IIP_3	Input-referred IP_3
IIR	Infinite impulse response
IL	Implementation loss
IM	Intermodulation
IM_3	Third-order intermodulation
IO	Input/output communication
IP_3	Third-order interception point
I/Q	In-phase/quadrature
ISI	Intersymbol interference
ISM	Industrial, Scientific and Medical radio bands
ISSR	Interferer suppression and signal reconstruction
ITU	International Telecommunication Union
ITU-T	ITU Telecommunication Standardization Sector
L1	GPS L1 frequency band (1,575.42 MHz)
LAN	Local area network
LNA	Low noise amplifier
LO	Local oscillator
LOS	Line-of-sight
LPF	Low-pass filter
LQFP-32	Low-profile quad flat package (32 leads)
LTI	Linear time-invariant
LVDS	Low-voltage differential signalling
MAC	Media access control layer of the OSI model
MODEM	Modulator–demodulator
MOS	Metal oxide semiconductor
MP	Multipath
MRC	Maximum ratio combining
MS	Mobile station
MSB	Most significant bit
MSE	Mean square error
MSED	Minimum squared euclidean distance
MSK	Minimum shift keying
N-FM	Narrowband FM
NBI	Narrowband interference
NLOS	Non-line-of-sight
NMOS	n-channel MOS transistor
OFDM	Orthogonal frequency division multiplexing
OIP_3	Output-referred IP_3
OL	Open-loop
OSI	Open Systems Interconnection basic reference model
OTA	Operational transconductance amplifier

PAE	Power-added efficiency
PAPR	Peak-to-average power ratio
PDF	Probability density function
PDP	Power delay profile
PHY	Physical layer of the OSI model
PL	Path loss
PL_{d0}	Path loss at d_0 meter distance
PLL	Phase locked loop
PMOS	p-channel MOS transistor
POTS	Plain old telephone service
PPM	Pulse position modulation
PRN	Pseudorandom noise
PSD	Power spectral density
PSK	Phase shift keying
PSO	Particle swarm optimization
PSTN	Public switched telephone network
QAM	Quadrature amplitude modulation
QPSK	Quadrature phase shift keying
rake	Rake receiver: a radio receiver using several sub-receivers
RDS	Radio Data Service (on FM 57 kHz subcarrier)
RF	Radio frequency
RFID	Radio-frequency identification
RMS	Root mean square
RPE-LTP	Regular Pulse Excitation with Long-Term Prediction
RS-232	IEEE recommended standard 232 for serial interfacing
RX	Receive
RZ	Return-to-zero (related to line codes)
SAW	Surface acoustic wave
SDC	Selection diversity combining
SDR	Software defined radio
SIFS	Short interframe spacing (IEEE802.11 related)
SNR	Signal-to-noise ratio
SoC	System-on-a-chip
TCM	Trellis coded modulation
TCP/IP	Transmission control protocol/Internet protocol
TF	Transfer function
THD	Total harmonic distortion
TSP	True single phase (related to dynamic logic circuits)
TDMA	Time division multiple access
TX	Transmit
UMTS	Universal Mobile Telecommunications System (3G)
USB	Universal Serial Bus

UWB	Ultra-wideband
VGA	Variable gain amplifier
Wi-Fi	Wireless Fidelity
WLAN	Wireless local area network

Symbols and Quantities

$\langle\cdot\rangle$	Arithmetic mean operator
$\otimes$	Convolution symbol
A	Frequency-independent amplification factor
A_0	DC-gain of $A(f)$
$A(f)$	Frequency-dependent amplification factor
B_c	Coherence bandwidth
BW_{3dB}	−3 dB bandwidth
C	Shannon capacity of a channel
C_{in}	Input capacitance
f_D	Doppler spread
dB	Decibel
dBr	dB referenced to some nominal level
dBV	dB referenced to one volt
dBW	dB referenced to one watt
ϵ	Error variable
$[\epsilon]$	Error vector
ϵ_{ox}	Dielectric constant of the gate oxide layer
$E[\cdot]$	Expectation operator
η	Efficiency
F	Noise factor
f_{LO}	Local Oscillator frequency
f_{max}	Maximum frequency of oscillation (see p. 195)
f_T	Unity current gain cut-off frequency
G	Gain factor
G_{ISSR}	Coding gain of the ISSR algorithm
g_m	Small-signal transconductance gain of a MOS transistor
H	Henry (SI unit for inductance)
H	Feedback factor in a closed-loop system
H(s)	Continuous-time transfer function (s-plane)
H(z)	Discrete-time transfer function (z-plane)
i	Small-signal AC current
I	DC current
i_{od}	Small-signal differential output current
J	Joule

k	Boltzmann constant (1.38e − 23 J/K)
K	Kelvin
K'_n	Intrinsic transconductance parameter for an NMOS (KP in Spice)
kTB	Thermal noise floor (kT = N_0, B = bandwidth)
λ	Wavelength
L	Effective gate length of a MOS transistor
L_{min}	Minimal gate length of a CMOS technology
μ	Mean
N_0	Thermal noise density ($N_0 = kT = -174$ dBm/Hz @ T = 290 K)
N_c	Colored noise density
NF	Noise figure
P_{avg}	Average signal power
P_e	Probability of error
Q-factor	Quality factor of a resonant system
Q(z)	Q-function (see p. 57)
r	Rolloff factor of a raised-cosine filter
R	Coding rate
r_{DS}	Small-signal drain-source resistance
r_o	Small-signal output resistance
R_s	Series resistor
S_{11}	Input reflection coefficient
S_{12}	Reverse transmission coefficient
S_{21}	Forward transmission coefficient
S_{22}	Output reflection coefficient
σ	Standard deviation
σ_τ	see τ_{RMS}
τ	Propagation time
τ_{RMS}	RMS delay spread
T	Temperature (room temperature = 300 K)
T_c	Coherence time
t_{ox}	Oxide thickness
T_s	Symbol duration
V_{GS}	Gate-source voltage
V_{GST}	Overdrive gate-source voltage above V_{th}
v_{id}	Small-signal differential input voltage swing
v_{od}	Small-signal differential output voltage swing
v_{ref}	Reference voltage
V_{th}	Threshold voltage
W	Effective gate width of a MOS transistor
ω	Angular frequency [rad/s]

ω_0	Angular frequency of the fundamental component
ω_{3dB}	−3dB angular pole frequency
$\omega_{p/z,n}$	Angular frequency of the *n*-th pole or zero
Z_{out}	Output impedance

Chapter 1

DIGITAL COMMUNICATIONS OVER ANALOG CHANNELS

It is well known that cost efficiency is one of the driving forces behind on-chip integration of CMOS digital circuits. The vast amount of useful application contexts combined with a potential for cheap high-volume production easily compensates for the high engineering and start-up costs. This unique blend has made integrated circuits one of the most important developments in the previous century and one of the driving forces behind today's economy. In the early 1990s, the speed and density of on-chip digital circuit elements achieved a critical mass which has led to the rise of the so-called *digital wireless communication*. Before this time, the center-of-gravity of wireless communications was almost completely slanted in favor of analog circuits. It is not suggested that wireless communication was limited to the transmission of analog data, though. Transmission of digital information was commonplace, but a major part of the signal processing was housed in the analog domain. Analog circuits were responsible for the signal preprocessing and a clean, refurbished signal was handed over to the digital back-end. The back-end itself, however, did not actively contribute to the actual signal recovery process. At best, digital circuitry was used as the controller for non-time-critical tasks upstream in the physical layer of a transceiver system. This includes, for example, off-line error detection, error correction and the higher-level retransmission protocols. Real-time signal processing such as filtering, automatic frequency correction (AFC) and also frame synchronization were taken into account by the analog or mixed-signal circuit blocks.

W. Vereecken and M. Steyaert, *Ultra-Wideband Pulse-based Radio: Reliable Communication over a Wideband Channel*, Analog Circuits and Signal Processing Series,

Figure 1.1. *Left:* the first commercial modem (AT&T, 1962). *Right:* V.92 laptop modem, 56 kbit/s, 85 × 54 mm.

EXAMPLE: BELL 103 MODEM

A good example is the Bell 103 (V.21) modem system[1] (Figure 1.1). In the transmission section of this modem, 1s and 0s are translated in two separate tones (1,070 Hz and 1,270 Hz), which are sent over the telephone line. At the receiver side, a PLL-based demodulator attempts to track the frequency of the received FSK signal. The decoded signal is recovered at the output of the loop filter.

Digital processing was not present in these early digital communication systems, simply because it could not match the speed requirements at that time. It was only from the early 1990s, with the appearance of more, faster and cheaper digital circuits that the required processing power became commercially available. It is very important to recognize the true impact of this shift from analog towards digital processing. Whether or not digital circuits would be superior to their analog counterparts is not relevant for this discussion, but it is indisputable that digital computing power has brought signal processing to a higher level of abstraction. In a digital world, signal processing is no longer performed by projecting a mathematical model on an analog circuit implementation. Digital signal processing (DSP) involves that the signal itself is virtualized and brought to the same abstraction level as the mathematical description of the system. The result is a much higher flexibility, not only because more complex

[1] Running at the mind-boggling speed of 300 bit/s [Swe05].

manipulations can be performed on numerical data, but also because software algorithms are not necessarily hard-coded in the firmware of a system. They can be updated and modified if needed.[2]

All those developments have recently led to the concept of *software defined radio* (SDR). In an SDR-based system, the role of analog circuits is pushed into background, ideally up to the point where the analog-to-digital (AD) converter is directly connected to the antenna.

Given the fact that – at least at the time of writing – there are no widespread implementations of this ideal concept, it is certainly questionable whether it is a good idea to disown, ignore, and utterly forsake analog circuits and move over to an all-digital implementation. The reason must be looked for in the distinction between the characteristics of the baseband signal and the RF waveform being transmitted over the wireless channel. The search for insight leads us to the baseband-RF interface and, more specific, its position in the transceiver chain. The baseband-to-RF transformation can be explained in two different ways, depending on the actual conversion process. In either case however, the goal is to reshape baseband analog or digital information in such a way that it can be transmitted in a specific frequency band over the air interface.[3]

In the first approach, information is directly attached to the RF-carrier. It is achieved by modifying one or more properties of the RF-carrier over time, hence the name (de)modulation. Some of the most familiar examples are on–off keying (e.g. morse code), amplitude modulation (AM broadcasting, video channel of analog TV) and frequency modulation (FM radio, audio channel of analog TV). One way or another, all these modulation techniques 'modulate' the carrier by changing the amplitude, phase (frequency) or a combination of both. It should be kept in mind that the above-mentioned modulation methods stem from the era of radio receivers where a crack in the glass of the second IF vacuum tube was more of a concern to electrical engineers than, say, bandwidth efficiency. Every modulation type was developed with a feasible analog implementation of the (de)modulator in mind. A noteworthy observation is that there is only little freedom to control the bandwidth usage of such directly modulated systems.

For example, in case of AM radio, variations in the amplitude of the analog audio signal are directly translated to amplitude variations of the RF-carrier. The bandwidth usage of AM radio – twice the bandwidth of the baseband audio signal – is merely a secondary consequence of the modulation type. It cannot be

[2] In the assumption of a general-purpose processor. Due to speed or power reasons, sometimes dedicated (non-programmable) functional blocks are required. They are generally implemented as high performance extensions of the capabilities of the general-purpose processor.

[3] In fact, this is frequency division multiple access (FDMA) in its most basic form. It allows multiple users to operate at the same time on the same physical wireless channel.

controlled independently from the amplitude information, despite some brave attempts that came in the form of single-sideband (AM-SSB) and vestigial side-band modulation (AM-VSB). Another ubiquitous example is FM radio, where the analog audio signal modulates the frequency of the carrier. The amplitude of the signal defines the degree in which the RF-carrier deviates from its center frequency. The amplitude of the carrier signal, however, maintains a constant level.

At first sight it might look as if FM has better control over the characteristics of the transmitted signal. After all, the bandwidth of the transmitted signal is defined by the modulation index,[4] while the amplitude of the carrier is fixed. The drawback of FM is that, although the total transmitted power level remains constant, there is no control over how this power will be spread over the spectrum. Furthermore, it is particularly interesting to see how the single analog (L+R audio) channel, for which FM broadcast was originally developed, was later on abused to accommodate a second audio channel (L-R) for stereo receivers (Figure 1.2) and again, about 30 years later, to incorporate a low-rate information data channel (RDS [CEN97]) at a 57 kHz subcarrier.

All those tricks and twists were brought into life in order to retain compatibility with older FM radio receivers. From these examples, it should be clear that before cheap digital processing made its entrance in the world of transceiver systems, modulation systems were optimized towards their implementation in analog hardware. It was only much later, when portable communications equipment became widespread, that the attention began to shift towards secondary parameters such as bandwidth efficiency or power consumption.

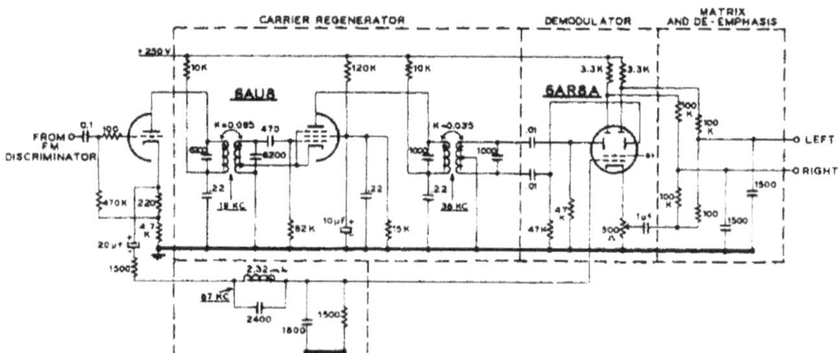

Figure 1.2. Stereo demultiplexer circuit with two vacuum tubes [DeV61] (Zenith Electronics Corp., 1961).

[4]The FM modulation index indicates the maximum deviation of the RF-carrier from the center frequency.

Today, one of the most vivid examples of the active participation of digital processing in the front-end of a transceiver is found in the GSM mobile phone system. The modulation method used in GSM is Gaussian minimum-shift keying (GMSK), a derivative of frequency-shift keying. The exact features of GMSK modulation will be withheld from this discussion, but the GSM system has a few remarkable details that are a perfect demonstration of the benefits of digital processing in wireless communications.

First of all, and most importantly, GSM is a true digital communication system. Starting from the very beginning, the intention of the GSM standard was to transport digital data. This in contrast to the analog-intended communication systems which were adapted later on, only to end up as the most horribly inefficient transport vehicles for digital data.[5]

In the GSM system, (Figure 1.3) things are turned upside down: the analog speech channel is immediately converted in a 13 kbit/s data stream by the analog-to-digital converter and the RPE-LTP speech codec [Eur90]. The biggest advantage of a digital representation is that the manipulation of data becomes very transparent: error correction information can be easily interleaved with the original raw data stream and data words can be reshaped into symbols of arbitrary size. The latter finding opens up a whole new realm of possibilities for the communications design engineer, who is no longer tied to the nature and quirks of the analog incoming signal. Before modulation, a fixed block of information bits can be restacked into new data *symbols*. Data can be spread over symbols vertically, by choosing the number of amplitude or phase/frequency quantization steps, and horizontally over a period of time,

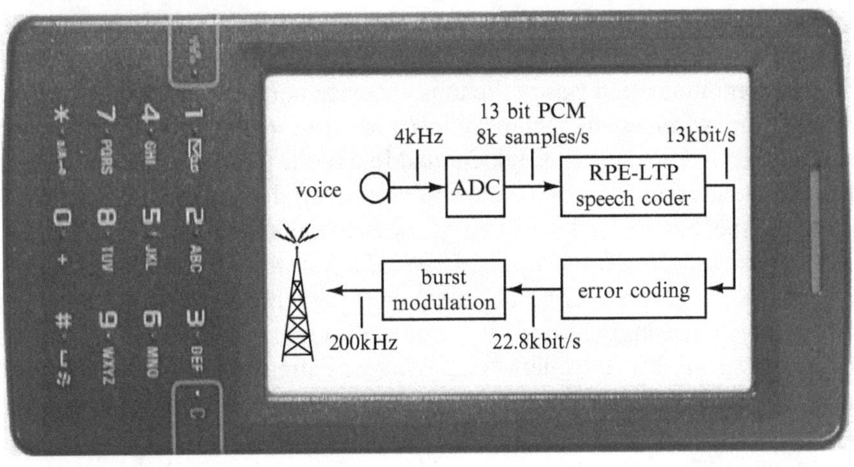

Figure 1.3. Manipulation of digital data in a GSM transceiver.

[5]Best possible proof: phone line dial-up modem.

by spreading the information over multiple symbols. Even better, not only the speed at which symbols are pushed through the modulator can be controlled, but also the smoothness in which the transition between two consecutive symbols takes place and even which particular symbol transitions are (or are not) allowed.

In the example of the GSM transceiver, digital information is flattened to a stream of single-bit symbols. These symbols are subsequently applied to a minimum-shift keyin (MSK) modulator.[6] The advantage of MSK is that it produces an *angle-modulated signal*, with a constant amplitude. This has a tremendous impact on the power consumption of the receiver: nonlinearities in the front-end of the transmitter will not result in in-band distortion. A nonlinear, but power efficient class-C amplifier can thus be used for constant-envelope signaling. In addition, the GSM modulator employs a very specific type of MSK, called Gaussian-shaped MSK (GMSK). Since the main spectral lobe of MSK is fairly wide, the incoming symbols are passed through a Gaussian pulse-shaping filter before they are applied to the modulator. The smoother transitions between symbols at the output of the filter results in a compaction of the power spectrum, and a more efficient use of the frequency bands allocated to the GSM operator.

1.1 Wideband radio: spectral and spatial efficiency

The previous findings bring us one step closer to the real topic of this chapter: having defeated the major shortcomings of pure analog modulation techniques, our attention can be narrowed to the real problem of radio communication: the limitations imposed by the wireless channel. Back in the good old days, most wireless communication was of the long-distance point-to-multipoint type. Today, the same wireless channel must host a multi-user environment with heavy, per-user traffic demands. It is evident that in a multi-user network, the number of simultaneous point-to-point connections in a certain area is at least as important as the spectral efficiency of a single device or the mobility of a single user. This concept is covered by the term *spatial efficiency*, which is expressed in units of bits per second per hertz per square meter (bits/s/Hz/m^2). Spatial efficiency is an increasingly important parameter for high-bandwidth data applications, such as wireless LAN networks. What was already being done at a larger scale, has now become common practice: the radio channel is spatially divided in increasingly smaller cells, allowing for a higher reuse rate of the scarce spectral resources offered by the wireless channel. For every multi-user application it is important to consider the trade-off between data rate and mobility.

[6] Minimum-shift keying is a subtype of continuous-phase FSK, where the frequency spacing between the two FSK frequencies is minimized [Pas79].

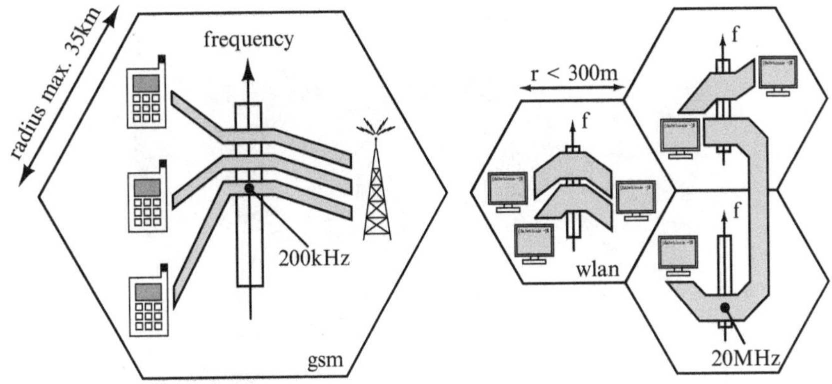

Figure 1.4. The GSM network has a high spectral efficiency and mobility. A WLAN system has higher throughput but a small cell size.

In case of the GSM network, for example, the emphasis is on mobility. In order to grant multiple concurrent users access to the GSM frequency band – each of them with considerable transmission power – spectral efficiency was one of the decisive factors in the choice of the modulation type. On the other hand, for WLAN systems, the emphasis is on throughput rather than on mobility (Figure 1.4).

According to Shannon's theorem,[7] increasing the capacity of any communication system requires either a proportional increase of the transmission bandwidth or an exponential increase in signal quality. For high-throughput networks, it is thus more beneficial to think in terms of channel bandwidth, rather than pursuing a higher transmission power. Moreover, increasing the transmission power is not a viable option because it would imply larger cell sizes and, as a consequence, less concurrent users.

1.2 Increasing the spectral bandwidth

It has become abundantly clear that a reduction of the cell radius and a large spectral bandwidth are the key enablers for high-throughput, multi-user personal network applications. At the time of writing, the wireless LAN[8] networks are the most successful illustration of this statement. Besides a decreased action radius, there are also a couple of less obvious disadvantages that come with wideband systems, though. As most readers will already be aware of, *free space* isn't exactly the most accurate approximation of the environment in which wireless applications are being operated. Whereas multipath delay

[7]Channel capacity = bandwidth × $\log_2(1 + \text{SNR})$ [Sha48].
[8]IEEE802.11a/b/g [Gas05] and Hiperlan/2 [Eur01].

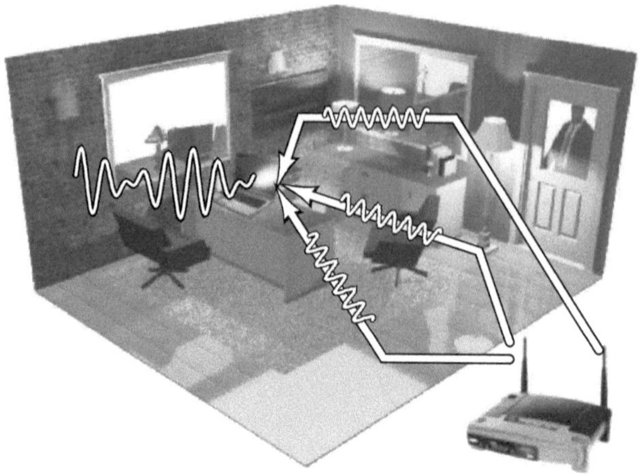

Figure 1.5. A multipath channel causes constructive and destructive interference. When consecutive symbols start to interfere with each other, the effect is called intersymbol interference (ISI).

spread results in a flat fading channel for most narrowband communication systems [Tse05], wideband communications suffer from frequency-selective fading in their frequency band. This means that some frequencies in the transmission band will be attenuated due to self-interference and other frequencies experience exactly the opposite effect due to the same mechanism.

From a time-domain point of view, fading is caused by multiple delayed copies of the transmitted symbol stream which arrive at the receive antenna. If the symbol period of the transmitted stream is smaller than the delay spread of the channel, this results in intersymbol interference (Figure 1.5). The problem becomes increasingly severe as the symbol rate goes up. Some of the more recent wireless LAN systems have abandoned the concept of a single carrier, exactly because of the intersymbol interference caused by a high symbol rate.

For example, the 802.11a standard uses a 52-carrier frequency division multiplexing system (OFDM [Eng02]), where each of the subcarriers is loaded with only a fraction of the total symbol rate. Thanks to the increased symbol duration, the susceptibility of a single subchannel to ISI is strongly reduced. Information is thus spread over multiple parallel data streams, which are transmitted in parallel over the frequency channels. If it happens that some of the subchannels become unavailable due to fading or narrowband interference, information can be redistributed over the remaining subchannels.[9] Of course, the data throughput of the transmission is still affected, but the bit error rate (BER)

[9] Adaptive loading of subchannels is not included in the current version of the IEEE 802.11a-1999 standard. All 52 subcarriers use the same modulation depth.

won't skyrocket as it does in a single-carrier system. Remark that the idea of using multiple parallel subcarriers has only recently found its way to a practical implementation. Thanks to the availability of abundant processing power, a discrete-time mathematical representation is used to build up the multicarrier baseband signal before it is converted to the analog domain.

Imagine how difficult it would be to generate and modulate 52 carriers in the analog domain, thereby controlling their frequency in such a way that these carriers are orthogonal to each other! After conversion to the analog domain, the only task that remains for the analog front-end is to upconvert the baseband signal to the appropriate frequency band. It seems that wireless communications have merely become a matter of digital processing, while the role of analog circuitry is ever more reduced.

However, in a world ruled by digital virtuality, it should not be forgotten that sooner or later, the digital processor has to interface with the analog world. Ironically, a nice example of the imminent danger of forgetting about the analog aspects is provided by the 802.11a/g wireless LAN standard itself. The combination of all the 52 QAM-modulated carriers will result in anything but a constant-envelope signal in the time domain. The sum of the amplitudes of all subcarriers results in a signal with a large peak-to-average power ratio (PAPR) [Han03]. An AM-modulated signal with a considerable modulation depth is thus applied to the analog front-end of the transceiver. To process such a signal, a very linear transmission chain with a considerable dynamic range is required. The dynamic range of the signal affects the loading factor of the DAC of the transmitter (ADC in the receiver). Since the average magnitude of the OFDM modulated baseband signal is less than its occasional peak value, the full quantization range of the converter is not used very efficiently. A higher number of bits is thus required to represent an OFDM signal with the same level of accuracy as a signal that uses the full-scale range of the converter.

The same situation is encountered in the RF power amplifier. The amplifier must be able to cope with the peak demands of the OFDM system, but most of the time it is driven at an average level which is considerably lower. For example, the PAPR of the 52-subcarrier 802.11a system requires that the amplifier is driven with a power back-off of at least 8 dB from the 1 dB compression point (P_{1dBc} [Hei73]). In other words, the RF power stage is being biased for operation under high-strain conditions, but handles much lower output levels for most of the time (see Figure 1.6). The reduction in the power-added efficiency (PAE) of the amplifier under such operating conditions has a direct impact on the power consumption, heat production and battery lifetime of portable devices.

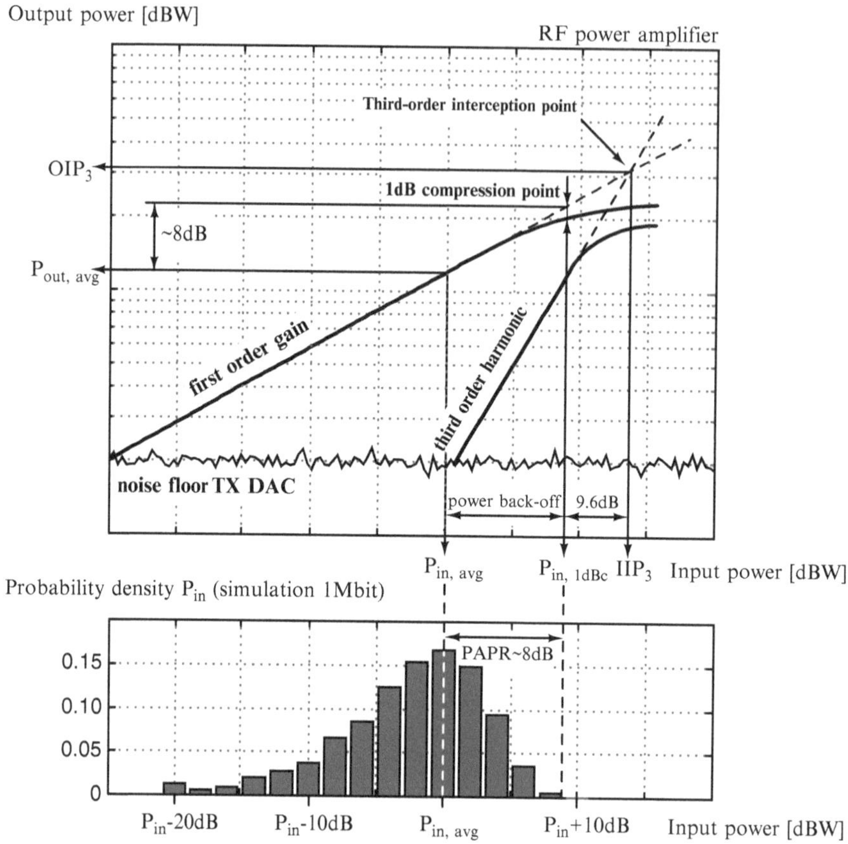

Figure 1.6. The *bottom* graph shows the power PDF of an WLAN transmission with 52 active subbands. The peak power of the RF amplifier (*top*) is 8 dB higher than the average output power.

This is also why, although the 802.11a standard was ratified in 1999, difficulties in the implementation of the 5 GHz analog front-end have delayed the availability of compatible products until 2003. In the mean time, the 802.11g standard was brought to life, as a transitional solution between the older 802.11b devices and the new 802.11a system.

Basically, 802.11g uses the same digital baseband subsystem as 802.11a, but the latter one is deployed in the 2.4 GHz band.[10]

The conclusion is that the changeover to a more intelligent baseband modulation algorithm (OFDM) has solved the issue of the non-ideal characteristics of

[10]Unfortunately, apart from offering much less available frequency space than the 5 GHz band, the 2.4 GHz band is also crowded with other users such as the older 802.11b WLAN network, Bluetooth devices and interference from microwave ovens [Kam97].

the wireless channel (i.e. ISI), but has opened at the same time a whole new world of challenges for the analog designer. During the design of a transceiver, none of the elements in the transmitter/receiver chain may be ignored or over-valued, whether it concerns the analog front-end, the digital back-end or the wireless channel itself.

1.3 Onwards to software defined radio?

Before continuing with the discussion of higher performing data networks, some reflection on the role of the analog front-end, the digital back-end and their mutual interaction is needed. First of all, it should be noted that our main goal is to transfer digital information from one location to another using a quite hostile and sometimes crowded interface: the wireless channel. The non-ideal properties of this information carrier include heavy signal attenuation, interference between co-cell users, self-interference due to multipath reflections and the frequency-dependent passband characteristic of the channel. It is our goal to make optimal use of the possibilities of both analog and digital techniques to convert the input signal from a representation of bits into a stream of electromagnetic symbols, dump this signal on the wireless interface and try to find out at the receiver side what the transmitter was trying to say in the first place. In the previous section it was already suggested that it is not always easy to make a justifiable choice for the modulation method or to find out where to draw the dividing line between the analog front-end and the digital back-end. The reconfigurability of digital hardware and the flexibility of software make digital signal processing especially suited for operation in a dynamically changing environment.

For example, remember the case of OFDM, where transmission over subbands that suffer from destructive fading could be avoided by rerouting the data stream over the remaining subchannels. Furthermore, using digital error correction techniques, the integrity of the received information can be guaranteed. Even analog information can benefit from a conversion to a discrete-time discrete-value representation: thanks to the regenerative character of digital information, noise does not accumulate over a digital transmission system. This raises the question whether the analog front-end still plays any significant role in the transmission chain. Of course, it is clear that sooner or later, the digital signal must leave the mathematical world in order to be applied to the antenna[11] as a real-world signal in analog form (i.e. using voltage, current, frequency, charge or *foobar* units).

There is something that deserves some extra attention, though. One part of the modulation process is almost never performed in the digital domain: the

[11] Antenna: impedance converter between our circuit and the wireless channel.

conversion of the baseband modulated signal to (and from) the RF carrier center frequency. This is due to the fact that analog circuits can be tuned to the frequency of interest, while maintaining a narrow operating bandwidth around this particular frequency. Apart from extra losses due to the non-idealities of the building blocks, tuning an analog circuit to a higher frequency has no direct impact on its power efficiency. For a digital system however, processing a baseband signal or the equivalent RF-passband signal *does* make a big difference. In order to handle the RF frequency components, the RF signal must be processed at Nyquist sampling rate. In a digital discrete-time sampled system, there is no equivalent to a tuned analog circuit. Of course the reader might argue that something like subsampling exists, but this technique basically comes down to baseband signal processing glued to a somewhat obfuscated frequency conversion step (frequency folding). To the knowledge of the author, there is no such thing as passband logic hardware.

In the transmitter, the conversion from a mathematical data representation to the analog signal that is applied to the antenna terminals is performed by the digital-to-analog converter (DAC) and some other building blocks such as the RF power amplifier. In the receiver, the modulated RF-signal with low fractional bandwidth is first shifted to a sufficiently low IF or baseband frequency and is then converted to the digital domain. Also noteworthy to mention here is that in the front-end of a receiver, out-of-band blocker filtering occurs before frequency conversion and amplification. By removing unwanted frequency bands early in the signal chain, the linearity requirements and power consumption of the remaining components are drastically reduced.

Due to the large dynamic range of the input signals that arrive at the receiver, this pre-filtering step is commonly accomplished by surface acoustic wave (SAW) filters [Har69]. This type of passive filters provides a sometimes impressive ratio between the passband insertion loss and the attenuation in stopband (typically more than 40 dB [Fre99]). The disadvantage of SAW filters is that they cannot be tuned to the desired frequency during operation, so they are typically designed only to block frequencies which are out of the frequency band of interest. In-band interferers are still present at the output of the preselection filter. This is why in a superheterodyne receiver architecture (Figure 1.7), the signal is first converted to a fixed intermediate frequency,[12] where it is filtered using a ceramic channel select filter (455 kHz/4.5 MHz/10.7 MHz). Finally, the IF-signal can be further converted to baseband.

Of course, all those frequency conversion steps and off-chip SAW filters do not make the superheterodyne receiver the most favourite architecture among

[12]The IF-frequency should be chosen high enough so that mirror frequencies fall out of the passband of the band select filter.

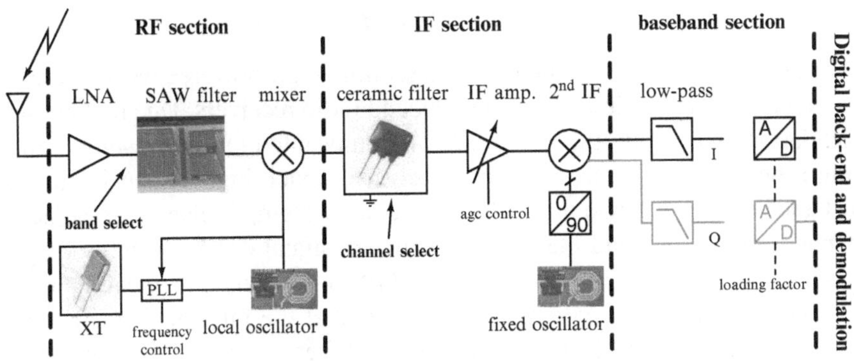

Figure 1.7. Architecture of a narrowband superheterodyne receiver.

followers of purely software defined radio. However, the previous considerations made clear why skipping those frequency conversion and filtering steps would impose serious demands on the linearity and dynamic range of the front-end in a purely software-based implementation.

Some receiver implementations have managed to get rid of the external components, though. In [Muh05] for example, the external SAW filter located at the IF frequency has been replaced by a direct-sampling mixer. The discrete-time continuous-amplitude output samples are subsequently processed by three on-chip IIR filters which are implemented using on-chip capacitor banks. A good linearity is still achieved since the IIR filters are located in front of the variable gain amplifier and the AD converter. The dynamic range of the IIR filter itself is much less of a problem since it largely depends on some passive components such as capacitors and CMOS switches. Doing the same in software, after the VGA and the AD converter, would place you one step ahead of your competitors concerning the reduction of power efficiency.

At the same time, the question is raised what to do with the ever-increasing bandwidth of telecommunication devices. In previous sections, it was found that for higher spatial efficiency, a small cell size should be combined with a large spectral bandwidth. From this point of view, the Shannon theorem indeed predicts that it is more advantageous to spread the total transmission power over a wide frequency band rather than to compress it in a small one. When bringing interferers into the game, this choice becomes much less evident: the large spectral footprint of the transmitted signal makes it a much more challenging task to remove an accidental in-band interferer. The poor selectivity inherent to wideband receivers, combined with a lower power spectral density (PSD) of the signal-of-interest, make that classic narrowband transceiver architectures are not the ideal candidate for porting to a wideband system.

1.4 Interference immunity issues of wideband radio

As briefly touched upon in the previous section, high-powered narrowband interference can be a serious issue in wideband radio receivers. Due to their large modulation bandwidth, the possibility of an unwanted interferer signal falling within the passband of a wideband receiver is much higher than for its narrowband counterpart. In a narrowband receiver system, such as GSM, adjacent-channel interference and the mirror frequency signal can be suppressed very effectively using the two-step RF-to-IF and IF-to-baseband conversion process. High-Q fixed-frequency ceramic filter sections at the intermediate frequency (IF) stage are very effective in suppressing blocker signals. When those filters are located in front of the variable gain amplifier and the AD converters, the linearity and dynamic range requirements of the subsequent circuitry is greatly relaxed which guarantees a good sensitivity of the receiver in the presence of strong blockers. In other words, narrowband receivers use their selectivity in the frequency domain as the principal weapon against interference.

Now let's do the same reasoning for a wideband receiver, for example the 802.11g system in the 2.4 GHz band. The best approach would be to remove in-band narrowband interferers before the VGA and AD converters. However, the number of interferers and their exact location in the passband of the receiver is unknown during design time. Suppressing unwanted signals in this way would require a series of tunable notch-filters in the analog front-end. It is clear that such a front-end would become impractically complex, so in practice the responsibility for a dynamically adjustable filter is deferred to the digital back-end.

Case study: OFDM modulation

The OFDM modulation method used by the 802.11a/g system provides opportunities for the transmitter to control how information is split up over the frequency spectrum and offers a more flexible replacement for the notch-filter bank in the analog domain.[13] Of course, all this comes at the expense of increased dynamic range requirements in the entire analog signal chain. In order to get some quantitative feeling for the impact of interference on the front-end of a receiver, consider the case where only one unintended narrowband interferer emerges at some location within the 802.11a/g channel. For the sake of clarity, the general outlines of the front-end of the OFDM receiver under consideration are first sketched below.

[13] Adaptive loading is not included in the 802.11a/g standard. For the reason behind this, see Section 4.2.

INDICATIVE FIGURES FOR AN 802.11A/G RECEIVER

Without proof, suppose that 10 bits are used as word length for the AD-converter in the 802.11a/g receiver. The loading factor of the converter [Tag07] is chosen to be −12 dB which means that, on average, only 8 out of the 10 bits are used to quantize the input signal. The main reason for limiting the RMS signal swing to about 25% of the full-scale range of the converter is the necessity to cope with the peak-to-average power ratio of the OFDM signal. Remember that, for the 52-subcarrier 802.11g standard, the peak PAPR of the time-domain signal is 17 dB. However, the possibility that all subcarriers constructively interfere is very low so a margin of only 8 dB – slightly more than 1 bit – should be sufficient.

The second reason for limiting the signal swing is to cope with the possibility of an incorrect gain setting in the variable gain amplifier (VGA) in front of the AD converter. In general, the automatic gain control (AGC) of the receiver attenuates the gain of the VGA in only a discrete number of steps. Rather than taking risk of clipping, which has a dramatic impact on the performance, the AD-converter is driven at some back-off value from the maximum signal level. Summarizing, 8 bits of the converter are used to obtain a sufficiently high SNR at the output and the 2 remaining most significant bits are only used to cope with peak signal values and the non-ideal behaviour of the AGC.

Now, suppose that the power of the narrowband interferer is of the same order of magnitude as the power of the OFDM transmission, say 20 dBm. If the distance between the (un)intended interferer and the OFDM receiver is larger than or equal to the link distance of the OFDM transmission itself, the input power as seen by the receiver will only rise by 3 dB or less. The headroom provided by the 2 most significant bits in the AD-converter is more than sufficient to handle this situation. The unwanted spur is removed from the spectrum in the digital back-end without further consequences on performance and life goes on. Then, the distance between the interferer and the receiver is halved. Using the free-space path loss model,[14] it follows that the interferer power available at the antenna of the OFDM receiver becomes four times (6.02 adB) larger. Most of the signal energy residing in the band-of-interest is useless, but clipping in the AD-converter is guaranteed if no further action is undertaken (see inset).

One of the possible measures to cope with the increased spurious power is to knit an extra bit on the word length of the converter. In best case scenario, doing so will double the power consumption of the AD-converter [Wal99], which is a

[14] $PL = PL_{d0} + 10\gamma \cdot \log_{10}(d/d0)$, with $\gamma = 2$(free space) $\cdots$ 4 and $PL_{d0} \approx 47$dB [Fri46].

hefty price for most portable applications. The other alternative is that for the same word length, an increasing number of bits is sacrificed to deal with the overhead of the narrowband spur. In any practical receiver implementation, this can be obtained by reducing the gain of the VGA. The consequence is that the noise floor in the unaffected frequency bands will rise, since the RMS level of the OFDM signal is now represented by a lower effective number of bits (ENOB). Actually, the interferer reduces the sensitivity of the receiver in all other subbands of the OFDM signal. The following example provides a rough estimation of the consequences for the data rate in an 802.11a/g link.

Effects of interference on the sensitivity of 802.11a/g

Each subband in the 802.11g system supports an M-ary QAM modulated subcarrier. The 802.11g standard [Wla07] specifies that the OFDM subcarriers can be modulated either using BPSK, QPSK, 16-QAM or 64-QAM (Figure 1.8). The total power allocated to a single subcarrier, however, remains constant over all modulation depths. This implies that for increasing modulation depths, the minimum squared euclidean distance (MSED) between the constellation points is reduced. Taking BPSK as the reference case (MSED = 1), the following Euclidean distances for the higher-order modulation depths are found (1.1):

EUCLIDEAN DISTANCES IN 802.11A/G

$$\begin{aligned} \text{BPSK} &: \text{MSED} = 1 = 0\text{dB (reference case)} \\ \text{QPSK} &: \text{MSED} = 1/\sqrt{2} = -3\text{dB} \\ \text{16-QAM} &: \text{MSED} = 1/\sqrt{10} = -10\text{dB} \\ \text{64-QAM} &: \text{MSED} = 1/\sqrt{42} = -16\text{dB} \end{aligned} \tag{1.1}$$

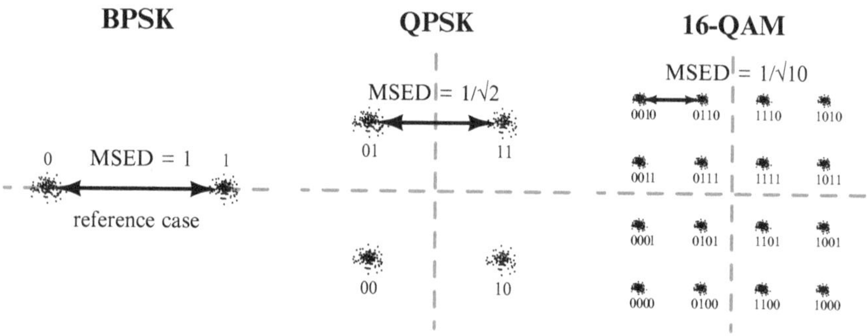

Figure 1.8. BPSK, QPSK and 16-QAM encoding maps for 802.11a/g. In each constellation, the MSED is normalized so that the average transmitted power is the same for all modulation depths.

If the noise power in a subband is increased, the magnitude-error on a demodulated QAM constellation point will also increase. In the constellation diagram, an error 'cloud' appears around the ideal constellation point (Figure 1.8). If the deviation from the original point becomes larger than half the distance between two constellation points, the (hard-decision) QAM-demapper of the receiver makes a wrong decision and decodes the incorrect symbol. Now consider the case where the SNR of the wireless channel is just enough to allow an 802.11a/g link to operate at its maximum data rate, being 54 Mbps using a 64-QAM modulation scheme [Wla07].

Suddenly, a narrowband interferer descends from the blue sky. Suppose that the power of this divine interferer is comparable to the transmission power of the OFDM link, but the distance between the source of interference and the receiver is half of the distance of the OFDM link. It was shown earlier that the receiver had to give up 1 bit of the accuracy of the AD-converter, only to be able to cope with the power of the spur. As a result of the loss of 1 bit, simulations show that the implementation loss (IL) due to the reduced signal-to-noise ratio is 0.37 dB ([Eng02] p. 129). Since the link quality was assumed barely enough to sustain the current data rate, the 802.11a/g link controller has three options: decrease the coding rate (R), reduce the modulation depth, or just fail.

The 802.11a/g standard uses two convolutional coding schemes at the 64-QAM modulation depth. For the maximum throughput of 54 Mbps, a coding rate of $R = 3/4$ provides a coding gain of 5.7 dB. If the coding rate is decreased to $R = 2/3$, the data rate drops to 48 Mbps in exchange for a coding gain of 6.0 dB. It should be clear that the additional coding gain of the latter case (0.3 dB) is not sufficient to compensate for the implementation loss of 0.37 dB. So the link controller is forced to take more aggressive measures to counter the reduced link quality: reduce the modulation depth. From (1.1), it follows that a fallback from 64-QAM to 16QAM increases the MSED by 6 dB. This means that up to four times more noise energy can be tolerated compared to the 64-QAM reference case, more than sufficient to compensate for the 0.37 dB implementation loss in the AD-converter of the receiver. The downside of all this is that the throughput of the 802.11a/g link is being reduced by more than 30%, from 54 to 36 Mbps, even for the fairly conservative case where the in-band interferer power is only four times larger than the average OFDM power.

The previous example shows that in-band interference has a serious impact on the sensitivity of a wideband receiver, if it cannot be properly removed from the signal chain in an early stage. In order to get some intuitive feeling for orders of magnitude, the interference immunity of OFDM-based wireless LAN is put in perspective to the blocking specifications of the GSM-900 standard (Figure 1.9).

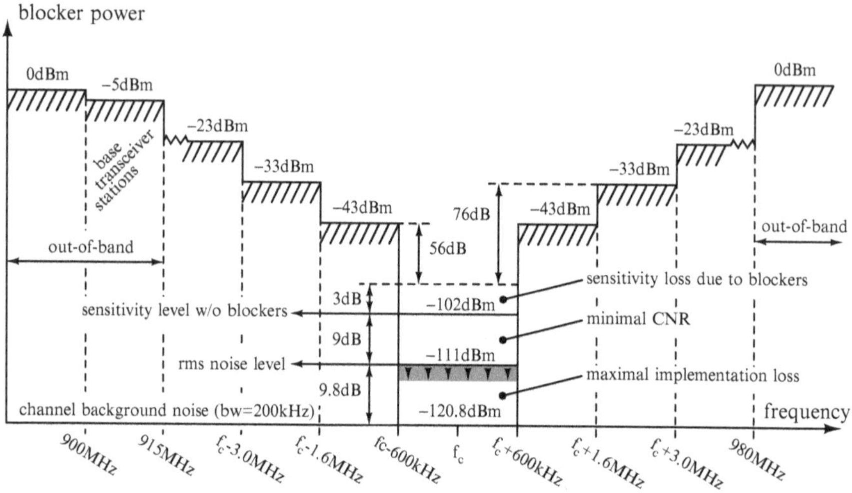

Figure 1.9. Blocking profile and link budget for a GSM-900 mobile station. Data was taken from the pan-European GSM 05.05 technical specifications [Eur05].

Unlike for the unlicensed 2.4 GHz band used by the 802.11g devices, the location of interferers in the GSM-900 frequency band is more or less predictable. The bandwidth of a GSM receiver is 180 kHz. The GSM system employs a very strict frequency allocation protocol and the interference from the first adjacent channel is separated by 600 kHz from the center frequency of the receiver. Specifications of the GSM systems require that the sensitivity of a GSM-compliant mobile station (MS) is better than −102 dBm. Furthermore, the blocking specifications state that this performance requirement is met for a useful signal that is 3 dB higher than the reference sensitivity level in combination with a blocker signal of −43 dBm at 600 kHz offset. The implementation loss (IL) is thus better than 3 dB, for a blocker signal that is up to 56 dB – more than five orders of magnitude! – higher than the signal-of-interest. In the imaginary case that the same in-band blocking specifications should be requested for an OFDM receiver, more than 10 extra bits would be required in the AD-converter to provide sufficient headroom. It is clear that the dream of a software-only radio receiver is not achievable in the foreseeable future.[15]

1.5 Organizational overview of this text

The work that is presented hereafter introduces a new wideband transceiver architecture, which is optimized for use in short-range high-speed data transfer applications. It is our purpose to armour the core of the system so that it is able

[15]Note: counting starts from zero in the year A.D. 2008.

to survive in a wide range of hostile channel conditions, such as in presence of in-band interference. Probably the most important aspect of a (wireless) communication system is the way the information is injected into the channel by the transmitter. During transport over the channel, noise and other unwanted artifacts will accumulate on the signal-of-interest, as a result of which a distorted image of the transmitted signal arrives at the receiver. It is the responsibility of the receiver to recover the original data, preferably with a low error frequency.

For this purpose, the receiver is commonly assisted by a combination of error coding and a clever modulation technique at transmission side. This is the topic of **Chapter 2**, which provides a high-level overview of the principles and use of coding and channel modulation in some widespread applications, such as the V.34 analog telephone modem and the successful wireless LAN networks (802.11a/b/g). One of the important lessons learned from these systems is that coding and modulation should not be considered as two separate processes in the transmission chain: depending on the mapping of information on the analog symbols in the modulator or the frequency-dependent transfer characteristic of the channel, certain bit positions in the transmitted data sequence may experience an increased probability of error. In a system ignoring this fact, the finite resources of the coding subsystem may be dispatched to the wrong location, resulting in a system performance that deviates from the theoretical optimum. Chapter 2 also introduces the notion of *coherence time*. In fast-varying channels, a short coherence time prevents the transmitter to adapt the transmitted signal to the current channel conditions. Chapter 2 concludes with a comparison between wideband single- and multicarrier systems, finding that single-carrier modulation techniques may not be completely ruled out in advance as they allow the use of an energy efficient, nonlinear power amplifier in the transmitter.

Chapter 3 pursues the same train of thought, showing that for a single-carrier modulated system in a frequency-selective fading channel, it is theoretically possible to cut away up to 40% of the frequency spectrum without any loss of information at receiver side. This is because in a single-carrier system, the energy that belongs to a single bit of information is automatically spread over the complete spectrum. This implies that, if some portion of the frequency spectrum goes missing for some reason (either due to fading or interference), the receiver must be able to recover the original data without intervention from transmission side.

In the second part of Chapter 3, the principle of *interferer suppression and signal reconstruction* (ISSR) is introduced. Running on the back-end signal processor of the receiver, the ISSR system is specifically aimed at the reconstruction of a QPSK-modulated signal with some missing frequency bands. It is also shown by simulation that ISSR, when combined with a Turbo-coder, is able to perform within 0.4 dB of the theoretical Shannon limit (Figure 3.7).

In **Chapter 4**, the foundations are laid for the main body of this work (Chapter 5 – Pulse-based wideband radio). A general overview is provided of the characteristics of a wireless transmission channel, including *power delay spread*, *frequency-selective* and *flat fading*, the *coherence bandwidth* and the *coherence time* of the wideband indoor channel. It is discussed here that a wideband transmission in a multipath environment suffers from flat fading, on top of the frequency-selective component. This flat fading component is a major issue for the reliability of the system, as it may cause periodic drop-outs of the communication link. The designer of a communications system may employ different techniques to prevent this from happening, which all rely on some form of diversity. In a spatial diversity scheme, for example, the incoming energy from multiple antennas is combined. If the signals that arrive at each of the antennas have independent fading characteristics, the probability that the combined signal experiences a deep fade is considerably reduced. A major drawback of the latter approach is that multiple antennas are an inefficient way to increase diversity, as it requires the duplication of costly (external) hardware.

With these caveats in mind, **Chapter 5** introduces the concept of pulse-based wideband radio. The advantage is that, for exactly the same symbol rate, short pulses provide a better multipath resolvability than a continuous-time modulated carrier. As a result, the receiver will be able to distinguish between a larger number of independently fading multipath components, which can be combined to form a more reliable link. What is repeatedly stressed throughout this chapter is that it is not a good idea to focus on the generation and detection of *individual pulses*. After all, the latter way of doing would require a perfect knowledge of the channel characteristics, which is a problem in a fast-varying environment (i.e. a channel with a short coherence time). Dealing with individual pulses does not allow for a straightforward solution to cope with intersymbol interference (ISI) or in-band interferers. Also, tracking individual pulses poses a serious synchronization issue to the receiver. For this reason, the proposed pulse-based radio system approaches the problem from a slightly different angle: instead of using pulses as a starting point, the pulse-based radio system that is introduced in Chapter 5 is founded on a basic QPSK radio subsystem. On top of this continuous-time QPSK sublayer comes a pulse-based extension layer. Although the transmitted signal has the appearance of a stream of individual pulses, the underlying system is still based on a regular QPSK transmission. As will be explained in more detail in Chapter 5, the advantage of this system is that it allows the QPSK subsystem in the receiver to employ a regular coherent detection scheme, without the need to be aware of the pulse-based extension shell.[16]

[16] Remember that the main function of the pulse-based extension layer is to increase multipath resolvability.

Chapter 6 ultimately puts the theoretical basis of the previous sections into practice and evaluates the prototype implementation of a pulse-based receive unit in 0.18 μm CMOS. The reader should keep in mind that the presented hardware is not a full-scale implementation of a stand-alone pulse-based radio receiver. The presented prototype contains the analog front-end of a single receive unit and is part of a larger, full-blown radio system, where multiple of these units work together in parallel. Chapter 6 goes through the general floorplan of the prototype chip, briefly discussing topics such as the RF input stage (which includes the pulse-based extension layer) and the baseband signal-processing circuits in the back-end. One of the major themes running through this chapter is the issue of spurious injection into the extremely sensitive signal-chain. This undesirable side-effect is caused by the presence of switching transient signals in the pulse-to-baseband conversion process. Chapter 6 also includes measurements results of the prototype chip, such as the aperture width of the pulse-based input stage, some test results showing the demodulation of a QPSK signal and a *noise figure* measurement on the complete front-end.

Chapter 7 elaborates in greater detail on a particular building block of the prototype receiver chip. The *variable gain amplifier* (VGA) in the baseband section of the pulse-based receiver employs an open-loop amplifier, which is loaded with a nonlinear output section. In a wide variety of applications, amplifiers are arranged in a feedback structure in order to suppress the second- and third-order distortion components caused by the nonlinear behaviour of the active gain element. Distortion suppression however, strongly depends on the amount of *excess gain* that is available in the feedback loop. Exactly the lack of excess gain at higher frequencies makes feedback amplifiers unsuitable for incorporation in a wideband signal chain. In contrast, the open-loop CMOS amplifier described here employs a nonlinear load, which allows to combine a good distortion suppression with a wide operating bandwidth. Chapter 7 provides both a rigorous theoretical background of the working principles and an in-depth mismatch sensitivity analysis of the nonlinear loaded open-loop amplifier. Again, the theoretical framework is supported by the implementation of an amplifier in 0.13 μm CMOS. Measurements show a flat frequency response in the 20–850 MHz frequency band and a 30 dB voltage gain. Under these conditions, the OIP_3 is better than 13 dBm over the aforementioned frequency band.

Some extra background on the frequency-dependency of distortion suppression in feedback amplifiers has been included in **Appendix 7**. The overview starts with the derivation of the well-known second- and third-order harmonic distortion formulas for a generalized amplifier in feedback configuration. After this, the expressions are extended in order to include the frequency-dependent behaviour of the active element. The calculations not only include second- and third-order distortion, but also account for feedback-induced third-order distortion, the latter one being caused by intermodulation beat products between the first-order signal component and second-order distortion at the summing node of the feedback loop. Armed with this knowledge, the maximum distortion suppression performance of a feedback system can be directly linked with the *cut-off frequency* (f_T) parameter of a particular CMOS technology. It is indicated that there is always a trade-off between gain, bandwidth and distortion performance in feedback-based amplifiers. Finally, taking the requirement of stability into account, the linearity performance limitations of the two-stage transistor amplifier in feedback configuration are derived.

Below is a brief schematic outline of the organization of this book.

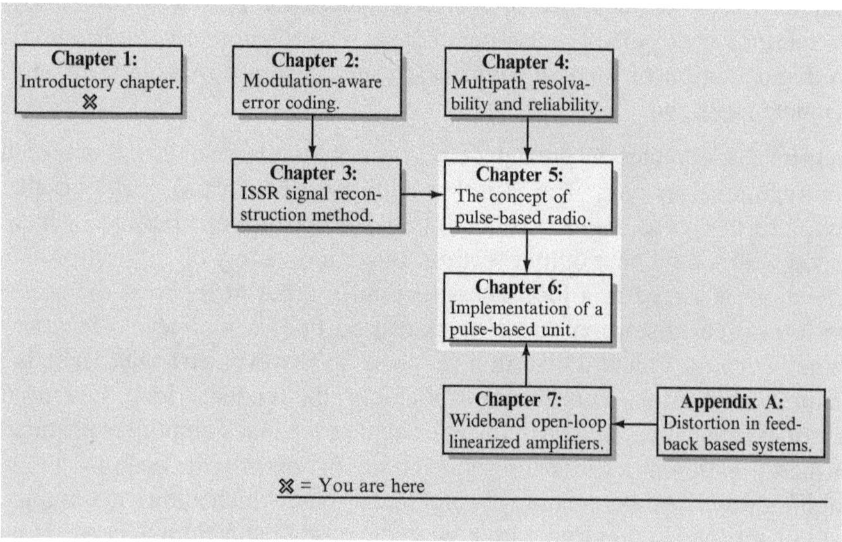

Chapter 2

MODULATION-AWARE ERROR CODING

In contrast to wireline networks, wireless communications are strictly resource-limited in every imaginable way. First of all, the wireless channel is a shared medium, which implies that its resources have to be split among multiple concurrent users (whether intended or not) at the same time and place. From a receiver's point of view, several unwanted signal components will be present at the antenna terminal, often several orders of magnitude stronger than the actual signal-of-interest. Even after aggressive filtering in the front-end of the receiver, a considerable amount of interferer power may still reside in the polished signal that is offered to the digital back-end. Furthermore, the transmitter is allowed to inject only a limited amount of power in the channel. All this, combined with conservative restrictions on bandwidth usage, make the wireless medium one of the most hostile environments for transporting information.

In the urge to get data as quick and as reliable as possible from one place to another, the efficiency of available resources is often pushed as far as possible towards the theoretical limit predicted by the Shannon theorem [Sha48]. Unfortunately, Shannon does not give any information on the way information should be attached to the channel, nor does it give any clue about the way to extract it again at receiver side. Consequently, chances are very likely that a transmission method which has proven its effectiveness in one application area will only make suboptimal use of the available resources if ported to another domain without further ado. This can be clarified by some simple examples.

For a start, it is important to recognize that both the analog front-end and the digital signal processor of a wireless transceiver are both located at the physical layer (PHY, layer 1) of the OSI model [Zim80]. The user of this PHY layer, in this case the data link layer (DLL, layer 2), expects a certain quality of the link.

W. Vereecken and M. Steyaert, *Ultra-Wideband Pulse-based Radio: Reliable Communication over a Wideband Channel*, Analog Circuits and Signal Processing Series,

For example, the specification of the 802.3 Ethernet DLL layer expects a BER better than $1/10^8$ [Eth05]. It is clear that this performance can simply not be met by an uncoded wireless link, since this would require an impractically high signal-to-noise ratio.[1] For this reason, the raw bit stream supplied by the data link layer to the physical layer (at transmission side) is always extended with redundant information by encoding the bit stream. Coding and error correction are commonly regarded upon as a purely digital matter, since it involves some decision making steps.

However, it does not imply that coding can be seen as a preprocessing step which is completely independent from the analog back-end. The error correcting capabilities offered by the coding algorithm does not come for free and strongly depend on the amount of surplus information that is appended to the original unencoded information stream. It is clear that a coding algorithm with a higher coding rate[2] will have the best performance in terms of BER. However, in a bandwidth-limited wireless channel, the extra throughput caused by coding cannot always be translated to an increased symbol[3] rate. In practice, there are two possible answers to the increased data rate as a result of the coding process: (1) reduce the data rate of the unencoded bit stream or (2) adapt the number of bits that are packed in a single symbol.

Unfortunately, there is no definite algorithm that provides a joint optimum for the bit rate, coding rate, symbol rate and the modulation depth[4] at once. Finding the (hopefully) optimal parameter settings is rather a process of trial and error, within boundaries set by implementation constraints in the analog domain (Figure 2.1). There are a few things, though, of which the designer of a wireless system should be fully aware. As a starting point, there is the Shannon theorem. From the available bandwidth and the expected signal-to-noise ratio, a good estimation can be obtained for the information capacity of an additive white Gaussian noise (AWGN) channel. If the bit rate of the unencoded data source is substantially lower than the maximum capacity predicted by Shannon, the channel is used below its capabilities. When the data rate provided by the DLL is above the theoretical capacity of the channel, it is obvious that things are predetermined to go terribly wrong, no matter how ingenious your combination of channel coding and modulation may be. In practice, the limited availability of computational resources in the decoder and restrictions imposed by the modulation scheme in the analog front-end require a certain back-off from the theoretical capacity of the channel.

[1] For BPSK, $P_e = \frac{1}{2}\mathrm{erfc}\left(\sqrt{\frac{E_b}{N_0}}\right)$. Thus, for a BER better than $1/10^8$, E_b/N_0 must be larger than 12 dB.

[2] Coding rate: the average number of output symbols (bits) used to encode one source symbol (bit).

[3] Symbol: a state or transition of phase, frequency and amplitude of the carrier which encodes one or more binary data bits. The symbol rate and the smoothness of the transition between two symbols determines the spectral footprint of the signal [Cou97].

[4] Modulation depth: defines the number of bits encoded on a single analog symbol. E.g. BPSK (1 bit/symbol), QPSK (2 bits/symbol), 8-PSK (3 bits/symbol), ...

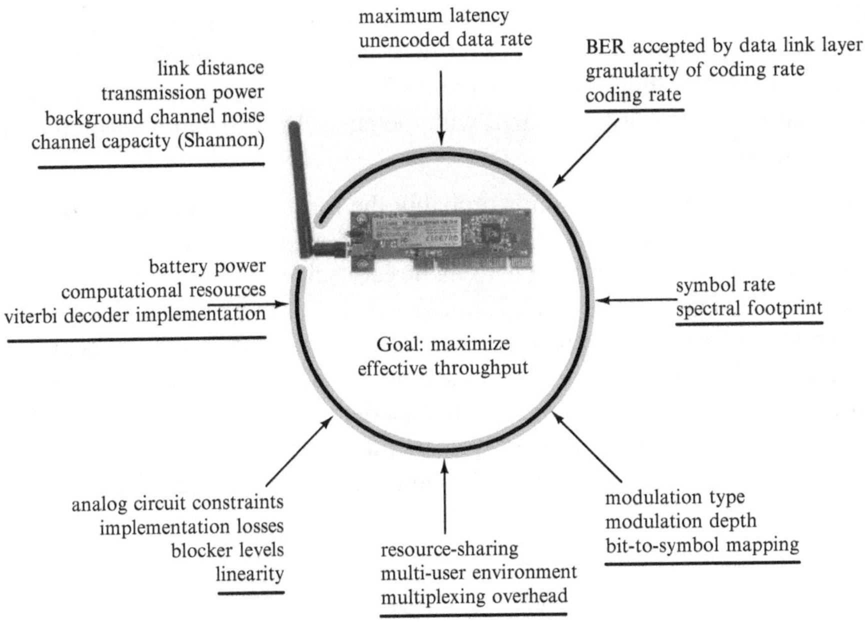

Figure 2.1. Trade-offs and constraints in the design of a wireless application. An ill-considered choice of one of these parameters results in a suboptimal use of the capabilities of the wireless channel.

At this moment, it is certainly interesting to note that the data rate of the encoded bit stream itself can be higher than the maximum throughput capabilities of the wireless channel. This observation can be explained by realizing that if the coding rate[5] is less than unity, each of the data bits produced by the coding algorithm contains less than 1 equivalent bit of unencoded 'information'. The remaining part is redundant information obtained from the same original data. As such, more encoded bits are injected into the channel, but the effective throughput of information always stays below the theoretical channel capacity.

2.1 Why error coding works

Remark that in most error coding schemes, it is not possible to relate a single input bit to a certain number of encoded bits. The encoding algorithm will spread the information of each unencoded bit over a longer sequence of encoded output bits. The length of the sequence over which the energy of one single input bit is being spread strongly depends on the number of internal states

[5]Typical values for coding rates are 1/2 (1 in, 2 out), 9/16, 2/3 and 3/4. A lower coding rate means that the output bit stream contains more redundant information.

of the encoder.[6] At the receiving end of the transmission chain, the decoder *despreads* the information embedded in a sequence of multiple data bits to a single information bit. The despreading effect of the decoder on the received bit stream is sometimes compared with the *energy despreading correlation* in a matched filter.

This spreading of information is probably the most important function of the encoding/decoding process. During transmission, noise accumulates on the transmitted symbols which leads to incorrect decisions of the symbol demapper. The recomposition of a single unencoded bit from a sequence of received data bits is a constructive process from an energy viewpoint, independent from the error coding strategy in use. On the other hand, the noise energy of a single bit error is spread out over multiple unencoded bits which are still embedded somewhere in the coded sequence, so that a single error partially affects the information of several information bits at the same time (Figure 2.2).

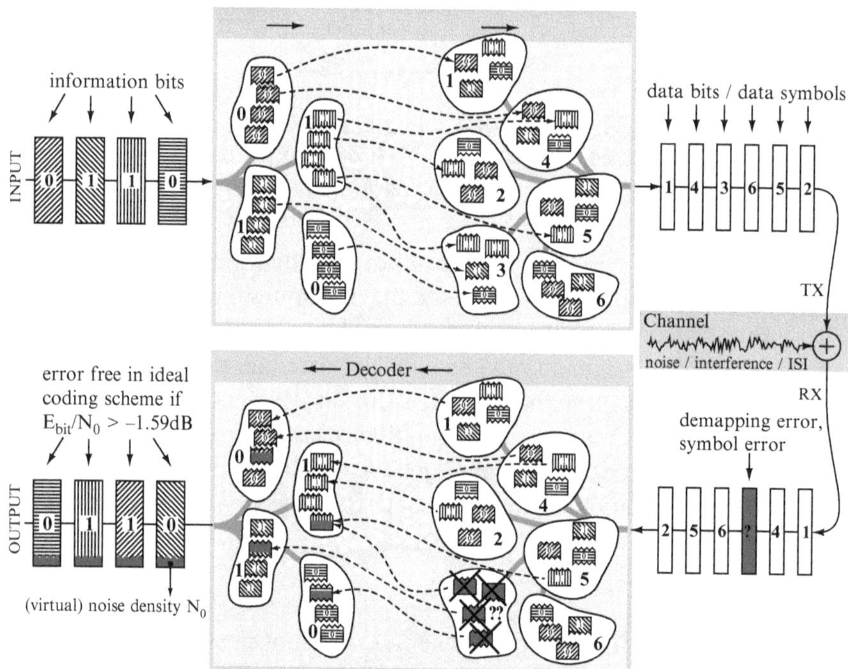

Figure 2.2. Error coding spreads the information associated to a certain input bit over multiple data symbols. If data is corrupted during transmission, the original information can be recovered thanks to redundant information in other symbols.

[6] A convolutional encoder can be expressed using a set of FIR filters [Rod95, Ros76].

When a data bit error occurs, the redundant information that is present in the unaffected part of the received data sequence comes into play. It is obvious that if no redundant information is available in the transmitted sequence (coding rate $R = 1/1$), even a single bit error will always result in an irreversible loss of information. Under such circumstances, there is little that the decoder can do to correct the mistake. For lower coding rates, a surplus amount of energy for each bit of unencoded information is embedded in the transmitted sequence. The availability of redundant information allows the decoder to detect and correct a limited number of errors. If the average virtual noise energy per information bit stays below a certain threshold[7] a well-built error coding algorithm should be able to successfully correct for a limited number of errors made by the symbol-to-bit demapper. Note the use of the term 'virtual noise': the whole process of deriving the original bits from the encoded bit sequence takes place in the finite-state-machine of a digital decoder. The actual form in which noise will manifest itself depends on the internal workings of such a decoder, but is not important for the general idea of error coding.

2.2 How error coding works

While the previous section has introduced some of the basic ideas behind error correction, it did not look deeper into the actual mechanisms that are used in the practical implementation of an error coding/correction algorithm. Rather than to give a comprehensive overview of the popular error correcting routines, it is the intention here to provide some insight in the underlying mechanisms that are common to most of these coding schemes. Two important characteristics always recur in a forward error coding (FEC) system. The first aspect is the presence of redundant information in some form in the encoded data stream. The second aspect is that the encoder spreads the information of a single unencoded input bit over a longer period of output data symbols. As a result, one data symbol of the encoded sequence may carry partial information of several input bits at the same time.

The length of the sequence over which information is spread depends on the type of the coding scheme. In the category of FE-block coding schemes, a fixed block of k input bits is transformed into a new block of l output bits. Since the memory of the encoder is limited in length to just one block of input bits, there is no dependency between two consecutive encoded blocks. Convolutional encoders are related to finite-impulse-response (FIR) filters: the encoder takes in a continuous stream of bits and produces multiple data streams [Rod95]. The internal states of the registers in the encoder are affected by each input bit. Every single input bit has an important influence on a complete series of subsequent

[7] The theoretical minimum E_b/N_0 value for error-free communication is -1.59 dB [Sha48, Cou97].

data bits at the output of the encoder. The duration of this influence strongly depends on the internal memory (i.e. number of possible register states) of the encoder.

The coding rate of the encoder is a crucial factor and gives the FE-coder its error correcting capabilities. If the coding rate is $R = k/l$, l encoded bits are produced for each k bits of information at the input. From this moment on, the reader is encouraged to reason in terms of bit-sequences instead of individual bits. For a certain sequence length, the resulting number of possible output data sequences is 2^{l-k} times larger than the number of bit combinations at the input. However, starting from a certain internal register state of the encoder, only a small *subset* of all possible output sequences can be produced by the FEC algorithm at any moment during the encoding process. Of course, the decoder at the other end of the transmission chain is more than fully aware of this fact.

Suppose that both encoder and decoder start with the same register states. During the transmission of the encoded sequence, noise accumulates on the signal. As a result, the demapper in the receiver makes an incorrect decision and some erroneous bits show up somewhere in the received data sequence. At the same moment, the decoding algorithm[8] in the receiver tracks the same path of states followed by the encoder and tries to recover the original 'unencoded' information. At the location of an invalid data bit, the received sequence shortly diverges from the path followed by the original sequence, and then remerges back to the correct path (Figure 2.3). The decoding algorithm, which is still tracking the succession of states followed by the encoder, perfectly knows the *allowed subset* of sequences that can be produced by the encoder when the registers are in a particular state. Therefore, the decoding algorithm tracks back from the moment that an error is detected and tries to recover the sequence with the largest possibility of being originally transmitted. This is obviously the sequence from the – at that moment – allowed subset which has the smallest possible divergence ('distance') from the received sequence. The probability of error depends on the minimum distance between the allowed sequences in a particular subset of the encoder.

This observation explains the nonlinear performance of error-coding techniques: When the sequence tracking algorithm in the decoder fails to recover the correct sequence, the internal registers will be trapped in an erroneous state. However, all future decisions of the decoder rely on the fact that the register states of encoder and decoder are synchronized. One single uncorrected error will propagate and generates a burst of consecutive errors.

[8]The most popular decoding algorithm used for FEC codes is the Viterbi algorithm [Vit67].

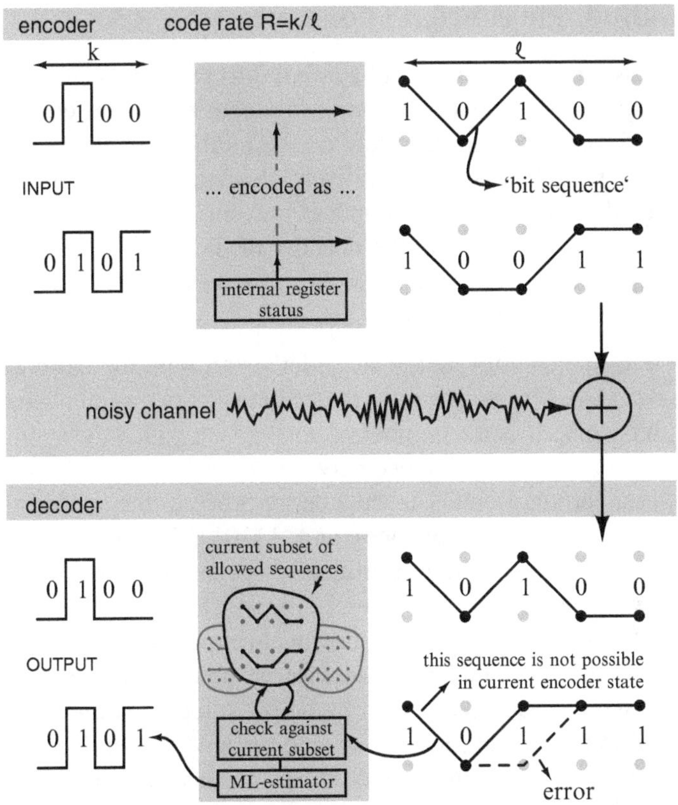

Figure 2.3. A number of bits at the input is encoded as a 'sequence' at the output of the encoder. At each moment, only a small subset of all sequences is in use, since the number of possible sequences is larger than the number of input bit combinations.

This also explains why coded systems have a so-called *coding threshold*. When the data error rate[9] drops below this threshold, the performance of a system with error coding becomes worse than the BER of an uncoded transmission. However, above the coding threshold, the error rate of coding will drop much faster than it is the case for an uncoded system. A good coding algorithm has a coding threshold which is as close as possible to Shannon's limit for error-free communication. For example, the original version of the high-performing *Turbo codes* [Ber93] shows a crossover point that is within 3 dB of the ideal error-correcting coding system.

[9]The coding threshold is sometimes expressed in terms of E_b/N_0. The conversion to the analog domain can be obtained using the BER versus E_b/N_0 relationship of the modulation system.

2.3 Coding: the concept of distance

In the previous section, the notion of symbol (bit) sequences and distances between sequences was introduced. It was also suggested that the error-correcting capabilities of a coding algorithm rely on the minimum distance between the sequences in a subset. It is also intuitively clear that in case of transmission errors, a large distance between sequences prevents that one sequence is accidentally mistaken for another in the decoder of the receiver. However, it was not explained what this 'distance between sequences' actually means and how this distance should be calculated.

In order not to overcomplicate the story, let's start with the basic example of *Hamming distance*. The Hamming distance between two symbol sequences of equal length is defined as the number of positions at which the symbols differ. For example, the Hamming distance between '010011' and '101011' is 3. Assume that, at a certain moment in the coding process this pair represents the complete subset of allowed sequences, and '111011' is offered to the decoder. Sequence '101011' has the largest probability of being transmitted by the encoder since its distance from the received sequence is only 1 bit. This is only true, of course, for as long as each bit has an equal probability of being flipped. If the sequence is transmitted over the wireless channel on a bit-by-bit basis, each data bit will be equally affected by noise. Another example is the information stored in the DRAM memory of a computer. Also in this case, there is no special reason why certain bits would have a sensitivity to failure which is different from the other bits.

The situation becomes completely different when some bit positions are systematically treated in a different way. It turns out that this is the case when bits are grouped and mapped on a constellation of symbols in the analog domain (Figure 2.4). Each of the 2^n points in such a constellation represents a collection of n bits (e.g. 3 bits in case of 8-PSK). When a symbol is sent through a noisy channel, 'clouds' will form around the constellation points when the received constellation map is visualized. It is more likely that the symbol-to-bit demapper in the receiver will confuse neighbouring points than two more distant constellation points.

EXAMPLE: 8-PSK WITH NATURAL SYMBOL MAPPING

Back to the example from above, where the encoder in the transmitter has produced the sequence '101011'. As is exemplified in Figure 2.4, suppose that this sequence is split in words of 3 bits ('101' and '011') by the 8-PSK symbol mapper in the transmitter. Remark that, in the case of natural mapping, the left-most bit (MSB) always controls two opposing constellation points, independently from the other two bits.

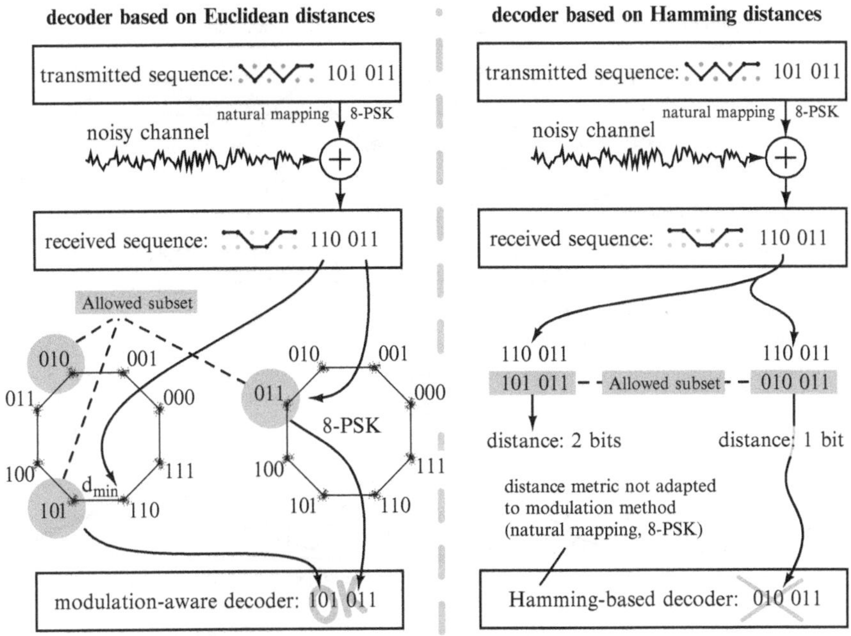

Figure 2.4. For 8-PSK modulation with natural symbol mapping, a distance metric other than Hamming must be used. Constellation points '101' and '110' have the shortest Euclidean distance, but have a Hamming distance of 2 bits.

As a consequence of the larger *Euclidean distance* between two opposing points in the 8-PSK constellation, the probability of error (P_e) of the MSB is obviously lower than for each of the other bits.

After transmission through a noisy channel, sequence '110011' is received. A decoder which is aware of the way in which the encoded data bits are modulated will correctly decide that '101011' has the highest probability of being the originally transmitted sequence. The Hamming distance based decoder would have chosen '010011'. This example explains the importance of correct symbol mapping.

The bit-to-symbol mapper method for a Hamming-based decoder should employ *Gray-coded mapping*, where the words of each adjacent constellation point only differ by 1 bit. Still, the Gray-coded symbol mapper does not correctly reflect the Euclidean distance between non-adjacent constellation points.

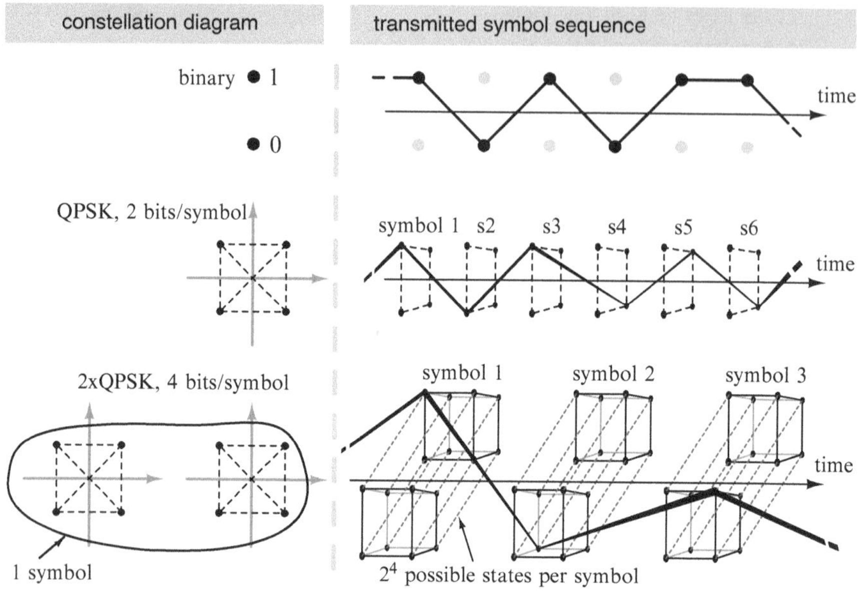

Figure 2.5. The output of the encoder can be mapped on multiple transmitted symbols. Higher-order dimensionality creates an interdependence between the symbols which can be used to improve the free minimum distance of the error coding scheme.

Instead of using the Hamming distance as a metric for the difference between two symbol sequences, the coding algorithm of modulated systems should thus rather employ the real *Euclidean distance* to compare corresponding points in the sequences. A simplified version of this concept was already given in the 8-PSK example above. However, the applicability of this idea is not limited to the case of a two-dimensional signal space. A sequence of encoded bits can also be mapped on multiple interdependent symbols at once, creating a higher order mapping space with its own specialized metric for the distance between constellation points (Figure 2.5). Higher-order mapping provides the encoder a more fine-grained choice of constellation points in a single (virtual) symbol representing a codeword, which results in a better spreading of the energy of the distance between sequences [Pie90, Lin04].

The first person who came up with these interesting insights in modulation-aware coding was Gottfried Ungerboeck, the inventor of *Trellis Coded Modulation* (TCM). Before the invention of TCM in 1982, the maximum data rate achieved by modems over the two-wire telephone lines (POTS) had been limited to 9,600 bit/s. This was still way below the theoretical limit of 34 kbit/s. The reason for this should be very clear by now: modulation and error coding at that time were considered as being two different disciplines. This is nicely illustrated by the ITU-T[10] V.2X modem recommendations [IT93]. Before version V.32, error correction was not even considered in the PHY layer of voiceband modems. Error control was separated from the modem's analog front-end and was defined in for example the V.42(bis) protocol recommendation. As a consequence, the error control mechanism was entirely located in the data link layer (DLL) of the OSI model, completely unaware of the far more intriguing activities that take place in the analog PHY layer of the modem.

2.4 Coding for a narrowband, noisy channel

In the previous sections, the basic principles of error coding have passed the critical eye of the reader.[11] One minor detail was not considered during these lively discussions, though. The error-correcting capabilities of coding are partially based on redundancy, which means that the encoded data rate is higher than the bit rate at the input of the encoder. If the capacity of the channel is has not been pushed to its limits yet, the excess data throughput can be compensated either by an increase of the symbol rate or by switching over to a higher modulation depth. However, when the channel is going to be used near the maximum theoretical capacity – which is almost always the goal in a wireless application – this implies that the rate of the encoded data stream will rise above the Shannon capacity of the channel. If the channel is bandwidth-limited and increasing the symbol rate is thus not an option, the only way to achieve this is to pack more data bits in each symbol (i.e. increase the modulation depth).

This method is used in the V.34 analog voiceband modems. The maximum symbol rate used by these modems is 3,429 Hz which, in combination with an average of 9.8 equivalent information bits per symbol, results in the well-known 33.6 kbit/s throughput rate. The bit error rate achieved by this system ($P_e < 1/10^5$) is fairly low, thanks to the use of a 4-D Trellis code variant [For96, Wei87]. Inevitably, the redundancy on which TCM is based will always be translated in an increase of the data throughput. The only way to cope with this larger throughput is to use a higher order modulation scheme. In the V.34 standard, a maximum of 1,664 constellation points or the equivalent of 10.7 encoded bits per symbol is used. The raw data rate through the channel, 36.7 kbit/s, is thus 7.9% larger than the information capacity of the PSTN line (34 kbit/s). As expected, this leads to a higher error rate in the encoded data stream. However, if the total energy per bit of information stays above the coding threshold of a particular coding algorithm, the net effect will be a decrease of the bit error rate at the output of the decoder!

[10] ITU-T: ITU Telecommunication Standardization Sector, better known as the former CCITT.

[11] To the unfaithful reader who skipped a few pages for some obscure reason: don't try this again.

One way to understand how this is possible is to consider once again the principles on which TCM is based: in the Trellis encoder, information bits are not sent on a bit-by-bit basis over the channel. Instead of this, bits are packed in sequences of encoded symbols. The information that belongs to each individual unencoded bit cannot be pinpointed to an exact location in this encoded sequence: information is spread over multiple data bits and symbols. The probability of error (P_e) when the receiver extracts the original bit stream from the encoded sequence is not dependent on the SNR of the individual encoded data bits, but on the minimum free distance between the allowed sequences.

Now here comes the crucial point in the understanding of modulation-aware error coding: by increasing the degree of redundancy, the distance between the constellation points is reduced because of the higher modulation depth. The SNR of the individual encoded data bits will decrease, but the minimum Euclidean distance between the allowed sequences is not affected by this. This is illustrated in Figure 2.6. On the other hand, it raises a new question: what is the benefit of increasing the degree of redundancy when it turns out that the distance between sequences remains unaltered? The answer can be formulated in two ways, but both approaches boil down to the same thing.

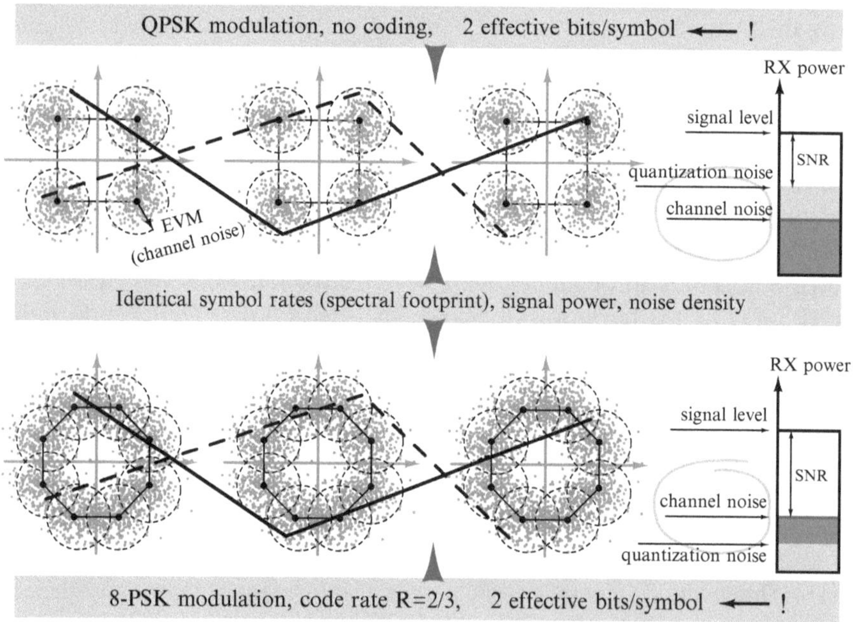

Figure 2.6. Coding allows to modulate below the noise floor of the channel, which rules out the quantization effects of the AD converter in the front-end. Despite the increased occurrence of symbol-to-bit demapping errors, the net result is a lower BER.

In one respect, a larger degree of redundancy provides a larger set of possible sequences from which the subset of allowed sequences is chosen. This allows for a more fine-grained control over the paths followed by each sequence and a smoother distribution of the mutual distance between the allowed sequences in a subset. After all, probability of error when extracting the original information from the encoded data stream is mainly determined by the weakest link, which is formed by the minimum Euclidean distance in every possible combination of two sequences in a subset.

Another interpretation is, that by employing a higher modulation depth, the accuracy of the modulation process falls below the noise floor of the analog channel. As a result, the detrimental effects of modulation in the transmitter and the quantization depth of the symbol-to-bit mapper in the receiver become negligible with respect to the noise floor of the transmission channel. Quite interesting to note here is that modulation and demodulation with accuracies below the noise floor of the channel is closely related to the ideas behind soft-decision viterbi decoding. The difference here is that the soft-decision process is already intrinsically supported by a TCM-based coding algorithm.

From this point of view, it should be clear that it makes no sense to use TCM at relatively high coding rates in a noisy channel: as soon as the effect of quantization noise becomes negligible compared to the channel noise, the benefits of further increasing the redundancy slow down rapidly. In addition, due to the large number of paths that have to be examined by the decoder, the power consumption of the receiver will increase, this without any significant return in terms of performance. Coding techniques merely based on Hamming-distances will even show a decrease in their performance: an increasing share of the total signal power is unnecessarily spilled in the encoding of distant transitions in the constellation diagram. However, this does not imply that TCM-based channel coding is superior. After all, TCM is a multilevel channel coding technique, while the algebraic coding schemes based on the Hamming distance metric can be of great value for bit-oriented applications, such as data storage systems.

2.5 Coding and modulation for a wideband channel: OFDM

It was already brought to the attention of the reader that coding is a crucial factor in the pursuit of an efficient and error-free communication link over a channel that is plagued by additive white Gaussian noise (AWGN). It was shown that for maximum (near-Shannon) throughput in a bandwidth-limited noisy channel, the coding process must be tweaked to the odds and the quirks of the non-ideal analog transmission medium. In the progress of the previous discussion, a few assumptions were made without further notice, since they are closely related to the nature of most narrowband channels. It was supposed that the power spectral density (PSD) of noise is stable over time and has a uniform characteristic over the entire frequency band. In most circumstances this is a fairly good approximation of reality, since many narrowband applications are designed to operate in their own, dedicated frequency band, safely separated from the interference caused by other users. This 'separation' can manifest itself under several forms. Apart from the physical separation such as in wired modem applications, communication channels can also be separated from each other using frequency selectivity (FDMA), time division multiple access (TDMA) or code division multiple access (CDMA). The common denominator to all these narrowband systems is that a high signal power is set against a low (in-band) thermal noise density. This high signal-to-noise ratio allows the use of a multilevel modulation scheme which is exploited by clever error coding algorithms such as Trellis coded modulation to achieve their near-Shannon throughput performance.

Wideband radio systems face a very different set of environmental circumstances, ensuing from the frequency-dependent characteristic of a wideband channel. First of all, there is the problem of frequency-selective fading in a multipath environment. From a frequency domain point of view, the receiver sees a signal-to-noise ratio which is heavily dependent of frequency. Some sections of the frequency band are completely unusable since the received signal power drops way below the noise floor. From a time domain point of view, a multipath channel suffers from intersymbol interference (ISI) if the period of the transmitted symbols becomes too short. Unfortunately, it turns out that high symbol rates are inherent to all broadband systems. Some modulation techniques, such as OFDM (802.11a/g), solve this problem by using multiple carriers. Each subcarrier is modulated at a reduced symbol rate, so that the resulting symbol period is below the delay spread of the multipath channel. OFDM also offers the possibility that the modulation depth of each subcarrier is controlled independently and is adapted to the current SNR as seen by the receiver in each of the subbands. This method can be regarded as a system-level *noise whitening* technique: by adapting the data throughput in each individual subband, it can be avoided that channels with a poor SNR determine the overall error rate of the system.

The main reason why adaptive loading is still not supported by the 802.11a/g standard is that the use of adaptive loading requires a feedback channel through which the receiver must inform the transmitter of the actual signal conditions in each subband. The major flaw of this method is that the transmitter must always work with outdated channel information. This is because of the delay between the actual channel estimation done by the receiver and the moment when the transmitter is informed of these new measurements and puts them to use [Tho00].

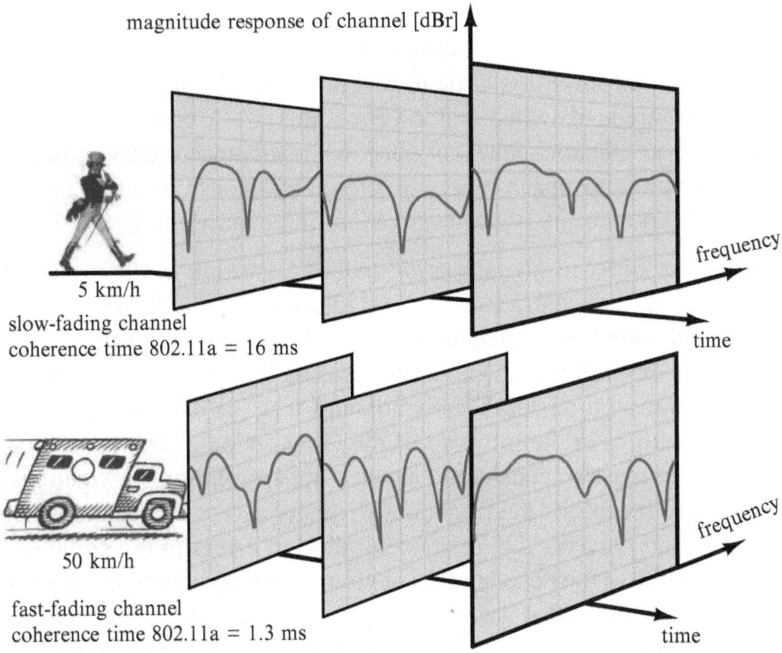

Figure 2.7. The short coherence time of a fast-varying channel prevents the use of adaptive bit-loading in OFDM-based systems.

This problem becomes increasingly serious in time-varying (fast fading) channels with a short *coherence time* as it will degrade the performance of the adaptive loading technique. As shown in Figure 2.7, the coherence time for a moving 802.11a device is between 16 ms (5 km/h) and 1.3 ms (50 km/h) [Sib02]. In the assumption that packets are transmitted without collisions between co-channel users, the minimum time delay between two consecutive transmitted packets from an 802.11a access point to a mobile client is 570 μs [Gas05]. The rate at which the transmitter gets updated on changing channel characteristics is thus sufficient for a slowly moving mobile station, but it is obvious that adaptive loading methods are predetermined to fail in fast-changing environments.

Remark that the current 802.11a/g standard implementations do not employ adaptive loading at all. The transmitter is not informed on the channel conditions experienced by the receiver and a single modulation depth is used over all OFDM subbands. In an attempt to gloss over the excessive bit error rates in some of the subbands, a channel equalization filter partially based on training data symbols is commonly used in practical OFDM receiver implementations.[12]

[12]In contrast to the transmitter, the receiver side is not standardized in the 802.11a/g system.

Unfortunately, equalization will never compensate for the low signal-to-noise ratios in subbands affected by destructive fading. Manufacturers of 802.11a/g-compatible devices solve this issue in several ways. The most drastic (but also effective) measure is to switch to a lower coding rate R or decrease the modulation depth over all subchannels at once. More advanced receiver implementations also exploit the space-diversity offered by the use of multiple antennas to diminish the possibility of destructive interference [Mat04].

Furthermore, in order to give the error coding algorithm better chances to recover the original message, a two-level bit interleaver [Cai98, Wla07] is placed between the convolutional encoder and the symbol mapper in the transmitter. Doing so, it is ensured that (1) consecutive data bits provided by the encoder are mapped onto non-adjacent subbands and that (2) data bits are placed at alternating locations in the bit word of the Gray-coded constellation diagram. Bit-interleaving must be considered as the stopgap solution for when all other techniques fail. It exploits the frequency diversity across the 52 subbands used in 802.11a/g and prevents a high concentration of errors in the stream of data bits that is applied to the decoder. It is obvious that bit-interleaving does not really solve the problems that are caused by the absence of the adaptive loading technique. It merely avoids a catastrophic failure of the error coding mechanism by equalizing the noise power over a longer sequence of bits. Obviously, bit-interleaving can and will never be a match for a decent adaptive loading subsystem. The conclusion here is that there is always a trade-off between complexity and power consumption at one side and implementation losses (IL) in terms of throughput at the other side. The quantity and nature of channel-related issues that are faced in the fast-varying wideband wireless environment have forced the IEEE 802.11a/g standardization committee to trade a significant amount of throughput performance for a more manageable hard- and software complexity.

2.6 Wideband single-carrier modulation

It has become abundantly clear that high-speed communication systems are not just an upscaled version of their narrowband counterparts. Multipath dispersion in the wideband channel opens a whole spectrum of new problems related to intersymbol interference (ISI) and frequency-selective fading. As if that's not enough, most broadband wireless applications with a high market potential are only licensed to operate in unlicensed frequency bands, crowded with a myriad of intended and not so intended interferers. At first sight, it would seem that every broadband system with a single-carrier modulation scheme is an easy victim on every front in such an hostile environment. Although it is perfectly possible to combat the problem of ISI with a time- or frequency

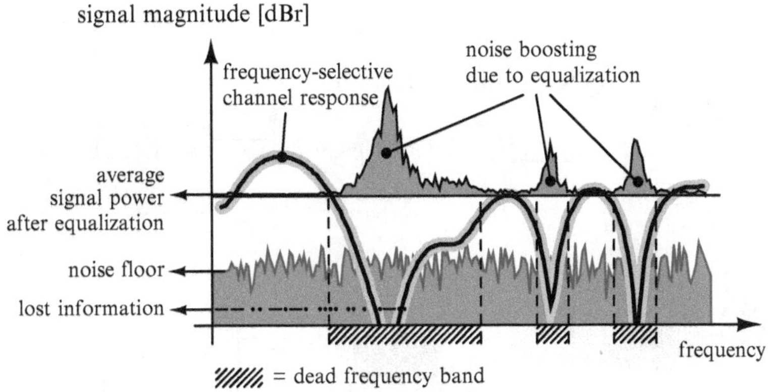

Figure 2.8. Channel equalization prevents ISI, but causes excessive noise when the channel response has spectral nulls (deep fades). Information in deep fades is submerged by the background noise of the channel.

domain equalization filter,[13] equalization may result in excessive noise boosting (Figure 2.8). This problem is inherent to equalization filters and is caused by deep fades in the frequency domain. After all, information which is lost during transmission cannot be recovered and is gone forever. From the point of error probability it can more advantageous to allow a limited amount of ISI and optimize the equalization filter for the smallest overall mean square error (MSE) ([Cou97] p. 184).

This seemingly innocent observation leads us to another, more problematic weakness of wideband single-carrier modulation: equalization offers no protection whatsoever against in-band interference. Even a single narrowband interferer occupying a small fraction of the total spectral footprint can (and will) bring down the performance of a whole wideband single-carrier system. Removing the interferer with a notch filter causes ISI and trying to remove the resulting ISI with a channel equalization filter introduces, again, a lot of noise.

Direct sequence spread spectrum

Despite all the drawbacks, there are wideband communication systems out there that stubbornly use a single-carrier modulation method. It is instructive to see how these systems cope with the difficulties of multipath and narrowband interference. Maybe the best known example from this category is the Global Positioning System (GPS). The encoding method used by the GPS system is *direct sequence spread spectrum* (DSSS).

---

[13]Equalization requires a cyclic prefix in order to deal with the causality of realizable filters.

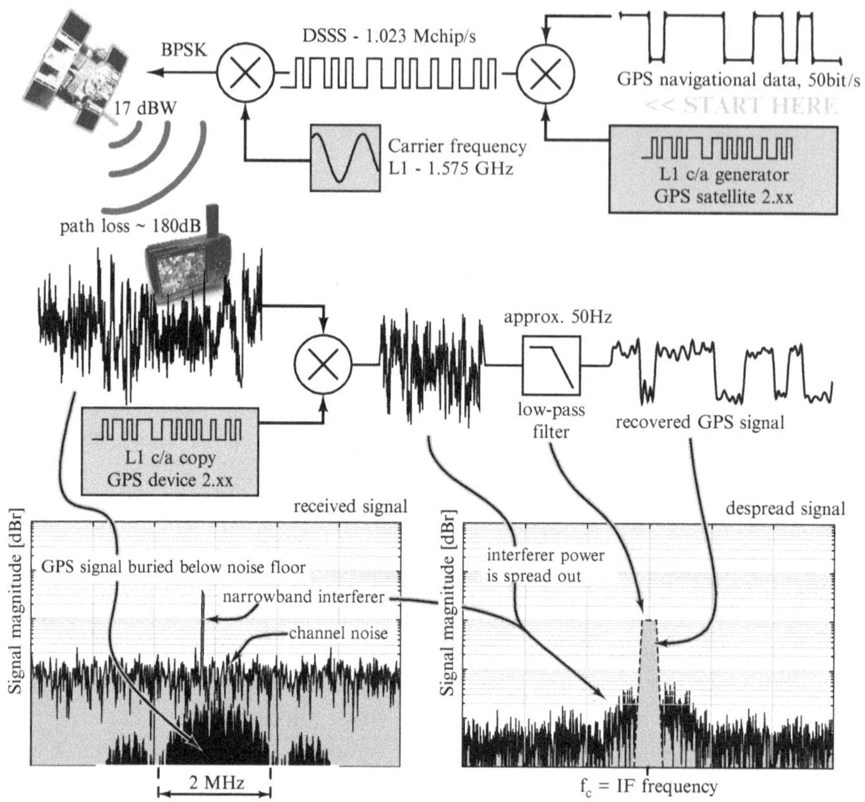

Figure 2.9. The GPS system uses direct sequence spread spectrum (DSSS) to extract the satellite signal from below the noise floor. The original information can be recovered thanks to the large minimum free distance in the subset of allowed code sequences.

DSSS is a channel coding technique where a low speed information bit stream is mixed (multiplied/XOR-ed) with a high-rate *spreading code*. This technique is known as frequency spreading, since the resulting signal at the output of the modulator has a much higher spectral footprint than the original information stream. In the receiver, a 'despreader' is used to extract the original low-frequency information from the spread data sequence (Figure 2.9).

In terms of signal-to-noise performance, the frequency spreading technique performs no better than a simple BPSK modulated system: the probability of error is only dependent on the integrated received bit-energy to AWGN spectral density (E_b/N_0) in the transmission band. However, DSSS offers a limited protection against accidental in-band spurious signals, because they are easily removed by the despreader in the receiver. The basic principles of despreading are very simple: mixing the spread sequence a second time with a replica of the spreading code will yield the original information stream. The effect of the despreader on a narrowband interferer is exactly the opposite: its energy is spread out over a large bandwidth and can be removed by a simple low-pass filter at the output of the despreader. In practice, things get a little bit more complicated: the offset of the spreading sequence generated locally in the receiver must be aligned properly to the spreading sequence in the encoded data stream. Failing to do so will result in a high-frequency noise signal, without a single trace of the original information. Interestingly enough, this property is at the same time the biggest disadvantage but also the biggest strength of DSSS-based systems.

This is best explained using the example of the GPS system. A low-rate 50 bits/s bit stream containing navigational information is spread using a 1,023-chip[14] pseudo-random spreading code which is repeated every 1 ms. The carrier of a GPS signal[15] is BPSK modulated at a symbol rate of 1.023 Mbit/s which results in a 2 MHz-wide null-to-null bandwidth of the GPS spectrum. Such a high symbol rate makes a GPS signal extremely vulnerable to ISI: the line-of-sight (LOS) signal between the satellite and the GPS receiver is easily disturbed by indirect multipath reflections. Intersymbol interference emerges when the reflected signal is delayed by one or more bit periods of the LOS signal. This occurs when the surplus path length traversed by the reflected signal is in the order of 300 m.

So how does the GPS system defend itself against multipath interference? One way or another, there must be a way for the receiver to *diversify* between the signal-of-interest and delayed versions of the same LOS signal. The solution lies in the characteristics of the spreading code used by the GPS system. The spreading code of a satellite is specially crafted so that it not only has a low cross-correlation with the codes belonging to other satellites, but also a very limited autocorrelation product. Only when the direct LOS signal of the satellite is multiplied by a correctly aligned spreading code, the signal is despread and the original 50 bits/s information stream becomes visible again. Delayed versions of the LOS signal are misaligned with respect to the despreading code in the receiver, as a result of which they are treated like any other unwanted interferer: the energy remains distributed over the complete 2 MHz spectrum. DSSS systems employ the time diversity of coded sequences to differentiate between signals and multipath reflections of the same signal. In other words, DSSS gives each GPS transmission some unique characteristics that allow it to be identified among all other signals, including a time-delayed version of itself.

[14] Data bits of the high-rate code are called *chips* to distinguish them from the low-rate information bits.
[15] C/A-code, $f_{center} = 1.575$ GHz.

Apart from its much higher level of redundancy, there is in fact little difference between DSSS and the traditional FEC-coding techniques. In case of DSSS, the sequences are controlled by the spreading code. A subset of only two complementary sequences is allowed at a certain moment, controlled by the alignment of the low-rate information stream with respect to the high-rate spreading code.

Also a very interesting observation here is that, at receiver side, the decoding of the received signal is performed before the demodulation. The reason for this is very simple, but has quite far-reaching consequences: It was already mentioned that the level of redundancy in the GPS system is higher than in most other wireless links. The power of the GPS signal seen at the antenna connector of a typical GPS receiver is in the order of −130 dBm. However, the total integrated noise power that is present in the same frequency band is no less than −111 dBm, considerably higher than the signal of interest. As a consequence, the GPS signal cannot be 'seen' by the receiver before it is despread, and it is thus also not possible to use the standard clock recovery strategies to create a stable time base for the demapper in the receiver.

Also, the coding threshold of most traditional decoder algorithms requires a bit-energy to noise power density (E_b/N_0) which is larger than unity. They will generally not be able to operate under the aforementioned circumstances. Because of this, the decoder of a DSSS receiver is commonly implemented as a despreading correlator. A correlating receiver, however, fails to function for as long as the correct alignment between the received signal and the despreading sequence has not been achieved. In fact, finding the correct timing offset of the despreading sequence is quite similar to achieving synchronization between the internal states of encoder and decoder in a traditional FEC error coding scheme. The difference is that synchronization in a DSSS system is usually not achieved in a single time frame. In practice, the receiver finds the correct offset of the spreading code using a more or less intelligent trial-and-error mechanism. The more processing power that is available, the more possible offsets the receiver can check out in a single time frame. This is of course at the cost of processing power and energy consumption. As a consequence, it can take a considerable amount of time before the receiver in a DSSS communication link is able to synchronize to the transmitter. From the moment the receiver is able to lock on the transmitted signal, the advantages of correlation processing become visible. In the example of the GPS system, the receiver won't even be able to detect the signal until correct synchronization has been achieved. Only after successful synchronization, the SNR is boosted by more than 43 dB as a result of which clock recovery and tracking becomes possible.

For moderate-length spreading sequences, the considerable synchronization delay for the receiver to lock on the transmitter makes the general DSSS system unsuitable for duplex communication links. This issue becomes especially problematic when a lot of short data packets must be exchanged, such as the TCP/IP packets in a WLAN network. Under these circumstances, only a very restricted spreading sequence length can be used.

A perfect example of this is provided by 802.11b WLAN devices. The DSSS modulation scheme employed in the 802.11b standard is *complementary code keying* (CCK). More details on CCK can be found in [And00], but the most important feature of CCK is that an 8-chip quadrature spreading code is used. From the total set of 4^8 (65k) possible spreading codes, a subset of only 64 spreading codes is retained. This allows the receiver to wrap 6 data bits in the 8-chip spreading sequence. The complete sequence can also be rotated by 0°, 90°, 180° or 270°. Using a differential encoding scheme, two extra bits are thus represented by the rotation angle of the sequence, which results in a total of 8 bits for each eight-chip codeword. Finally, the encoded data is sent at 11 Mchip/s through the channel. In the receiver, the channel is monitored by a bank of 64 parallel correlators and only the codeword with the largest correlation result is selected. It is indeed thanks to the large amount of parallel processing power that synchronization is achieved within the duration of a single symbol interval. If the subset of 64 codewords is carefully selected with respect to the auto- and cross-correlation properties, limited protection against multipath interference can be achieved. However, the practical coding gain of CCK is limited to only 2 dB above the performance of uncoded QPSK. This is due to the small minimum distance between neighbouring codewords [Hee01]. The resistance of CCK against multipath interference is very limited if compared to the results that can be achieved in the (low-rate) GPS system.

A severe disadvantage of the CCK system used by 802.11b compatible devices is the limited spreading factor. The consequence is that the energy of narrowband interferers is indeed averaged over the spectrum, but cannot be removed: after despreading (correlation), the complete frequency band is still used by the signal-of-interest. This is in sharp contrast to the spectral redundancy exploited in the GPS system. Chances that a 802.11b transmission survives the interference of a moderate powered in-band spur are thus very slim. But as bad luck would have it, 802.11b devices are only licensed to operate in the unlicensed 2.4 GHz ISM[16] band. This band was opened up by the FCC in 1985 for low-power unlicensed communication devices, only because that part of the spectrum was already polluted anyway by RF-leakage from applications such as industrial heaters and microwave ovens.

The RF-energy leaking from a microwave oven, for example, makes repeated sweeps from 2.4 to 2.45 GHz [Kam97], temporarily blanking out parts of the wireless channel for wireless devices using the same frequency band. To

[16] ISM band: internationally reserved radio band for industrial, scientific and medical use.

make things worse, multiple communication devices present at the same geographic location will try to claim the same frequency band. The MAC layer of 802.11-based WLAN's has anticipated on such occasions and provides a semi-orchestrated channel access method. The contention-based random access method (CSMA/CA[17]) gives all devices equal opportunities to acquire access to the wireless channel. Unfortunately, the same frequency spectrum is also in use by other competing networking technologies. The MAC layer of Bluetooth devices, for example, is kept simple and does not bother that much about the potential presence of other devices that operate in the same 2.4 GHz frequency band. The overall conclusion here is that DSSS modulation is not well suited for use in high-throughput wireless communication networks. The low spreading factor results in a relatively simple, but also a very vulnerable modulation method if deployed in a hostile multipath environment.

On the other hand, a single QPSK-modulated carrier has one competitive advantage over the OFDM modulation method discussed in Section 2.5: in a multicarrier system, subcarriers align in phase every once in a while, resulting in a high peak signal value. In a single-carrier system, the time-domain signal has a much lower PAPR value. As a result, the RF power amplifier of a 802.11b transmitter needs a much lower back-off value from the 1 dB compression point, so the PAE[18] will be considerably better than for 802.11a/g devices.

2.7 Conclusions on single- and multicarrier systems

As a conclusion to this section, some of the problems that must be tackled in a wideband wireless channel are summarized. Both the advantages and disadvantages of single- and multicarrier systems are considered here. From this, the properties of an ideal (fictive) transceiver system are determined.

First of all, multipath is a severe problem, since it can cause ISI for high symbol rates. In an OFDM-based system, multipath reflections are dealt with by splitting up the symbol stream over multiple carriers, each of which is modulated at only a fraction of the cumulative symbol rate. In a single-carrier system, some tolerance against multipath reflections can be obtained by exploiting diversity in one way or another. For example, in a DSSS-based architecture, diversity of coding is used to differentiate the signal-of-interest from unwanted reflections and interference.

Unfortunately, as pointed out above, the resistance of DSSS against frequency-selective fading and narrowband interference depends heavily on the frequency spreading factor. For high-throughput applications, the practically achievable

[17] CSMA/CA: carrier sense multiple access with collision avoidance.

[18] PAE: Power-added efficiency.

spreading factor is far too limited to obtain a satisfying tolerance against the imperfections of the wireless channel. By contrast, adaptive bit loading in OFDM offers much more opportunities to fight against multipath or interference, but requires active intervention from transmission side. Since the transmitter must be informed by the receiver of the current channel conditions, a feedback path from the receiver to the transmitter is needed. This can become problematic for channels with a low coherence time (stability) which explains why adaptive loading of the subcarriers is currently not included in the current OFDM wireless standards. Although WLAN devices supporting OFDM are often credited for their superior resistance against channel imperfections, in practice their lead is not as big as it might appear due to the absence of bit-loading. Alas, as a result of the widespread use and the success of these WLAN products, it has become extremely difficult to introduce fundamental changes in the system without losing compatibility with the existing legacy products.

The ideal transceiver system (Figure 2.10) would combine the advantages of both single- and multicarrier modulation. A single-carrier phase-modulation technique would be employed in the transmitter. This allows for a non-linear and thus energy efficient power amplifier. Furthermore, information should be transmitted using time and frequency diversity. The reason for this is that imperfections of the channel are almost always located in a confined area in time and/or frequency. By distributing the energy of unencoded information

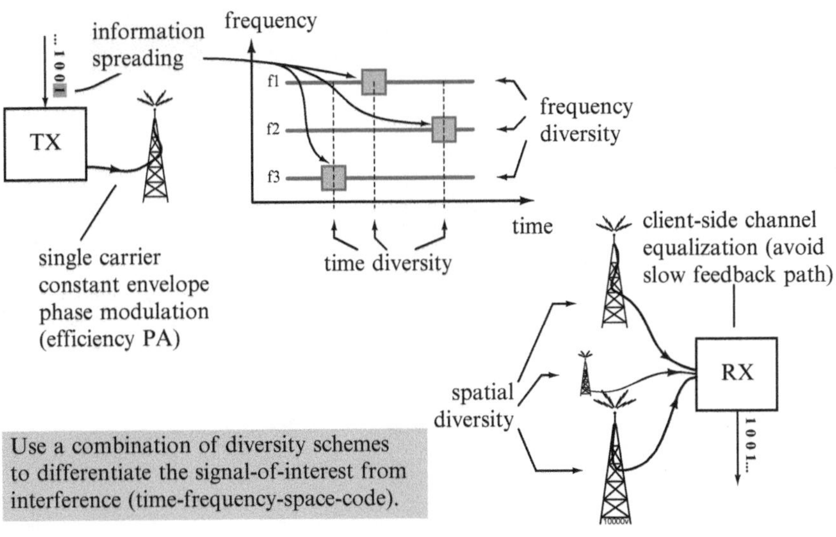

Figure 2.10. The ideal transceiver system would combine constant-envelope modulation with multiple diversity schemes in order to increase the reliability of the link.

bits over several dimensions (time/space/frequency), it is prevented that a temporary black-out of the channel corrupts a complete sequence of information bits. If a sufficient amount of energy is still available in adjacent dimensional planes, it should be possible to restore the original information sent through the channel. This is only possible, of course, when some redundant information has been made available by the transmitter.

Also, just like in the case of Trellis coded modulation (TCM), the redundant information should be attached to the data stream without affecting the information capacity of the system. At first sight, it may seem that these requirements are in direct conflict with the decision to use a single-carrier modulation method. However, that this is not necessarily always the case will be unveiled in the subsequent sections. And finally, last but not least, during the development of a new transceiver system, one must never forget the high inertia of bringing new technologies into the market. It must be possible to introduce the changes gradually and in a non-intrusive way in existing products, without breaking backward compatibility.

Chapter 3

MODULATION-AWARE DECODING: SIGNAL RECONSTRUCTION

The wideband nature of analog signals involved in high-speed digital communication systems requires a likewise high bandwidth of the transmission channel. Most wideband transmission mediums suffer from frequency-selectivity caused by reflections or multipath propagation. In addition, wideband wireless channels are increasingly vulnerable to imperfections such as time-varying channel-selective fading and in-band interference. In the previous sections, it was discussed that OFDM has been widely used to reduce intersymbol interference and that adaptive bit loading can be employed to avoid those parts of the spectrum that are contaminated by narrowband interference. All modulation techniques analyzed before try to preshape the transmitted signal is such a way that the receiver is able to extract the original information, even when the received symbol stream gets corrupted by the non-ideal transfer characteristics of the channel. For this reason, this type of channel coding techniques have been classified under the name *modulation-aware coding* systems. All these techniques have a common flaw, though: the characteristics of the channel are known to the receiver, but not at transmission side. A lot of the effort that goes into encoding, modulation and decoding gets lost due to inaccurate assumptions about the wireless channel. As a consequence, a significant portion of signal power must be allocated to the redundancy of the system, which is at the expense of the effective throughput of information. Also, the responsibility of adaptive channel coding is entirely allotted to the transmitter, so this type of approach is not very useful for a point-to-multipoint broadcast system.

This section describes an *interferer suppression and signal reconstruction* (ISSR) method for wideband communication systems. The ISSR technique shifts a big portion of the responsibility of dealing with non-idealities of the channel to the receiver. The signal reconstruction strategy used in ISSR should

W. Vereecken and M. Steyaert, *Ultra-Wideband Pulse-based Radio: Reliable Communication over a Wideband Channel*, Analog Circuits and Signal Processing Series,

not be confused with channel estimation and equalization. Equalization is focussed on the compensation of the frequency response of the channel. Instead of this, ISSR can decide to consider seriously corrupted frequency bands as irreversibly lost and subsequently tries to reconstruct the original data using the remaining, unaffected parts in the spectrum. On the other hand, ISSR is not a holy grail system and cannot serve as a replacement for channel equalization. ISSR is meant to be put in action when channel equalization fails due to insufficient signal quality in some of the subbands. The ability to deliberately ignore a specific part of the spectrum is also the biggest strength of ISSR. When a certain subband suffers from destructive fading, equalization will boost the gain for that particular band in an attempt to reduce ISI. However, the useful information within this band is often limited due to the background noise of the wireless channel. A lot of noise is thus inevitably introduced by the channel equalization filter.

The ISSR method is not willing to make this compromise between ISI and background noise. It will simply reject those subbands with an insufficient contribution to the overall signal quality (Figure 3.1). The rejection of subbands is not only a useful feature for a channel that is plagued by frequency-selective fading. The same feature can be thrown into the game to fight against interference in some part of the signal spectrum. For ISSR, it is not important to know exactly how a certain portion of the spectrum had gone missing. Rather the fact that it is missing in the first place is a crucial step in the signal reconstruction process of the ISSR method. Before jumping into the concrete specifics of ISSR, it is important to fully recognize that ISSR is running at the client-side of the wireless connection, hence it has the ability to react almost instantaneously to changes and unexpected events which take place in the channel. If correctly implemented, the performance of ISSR will be unmatched in comparison to server-side modulation-aware coding techniques such as OFDM.

3.1 Principles of signal reconstruction

One of the target applications of the ISSR system described in this chapter is the 802.11b wireless LAN system, the oldest of the three Wi-Fi standards. It was already mentioned in Section 2.6 that 802.11b data is encoded using DSSS, using the 11-bit Barker code or complementary code keying (CCK) as the frequency-spreading code [Con00]. Unfortunately, the effective spreading factor of these codes is substantially limited as a result of the considerable throughput of data (from 1 up to 11 Mbit/s) within a bandwidth of 22 MHz. As a consequence, the promises made by the hologram labels on the package of an 802.11b device are often not fulfilled by the device itself: frequency-selective fading in the indoor environment or in-band interference originating

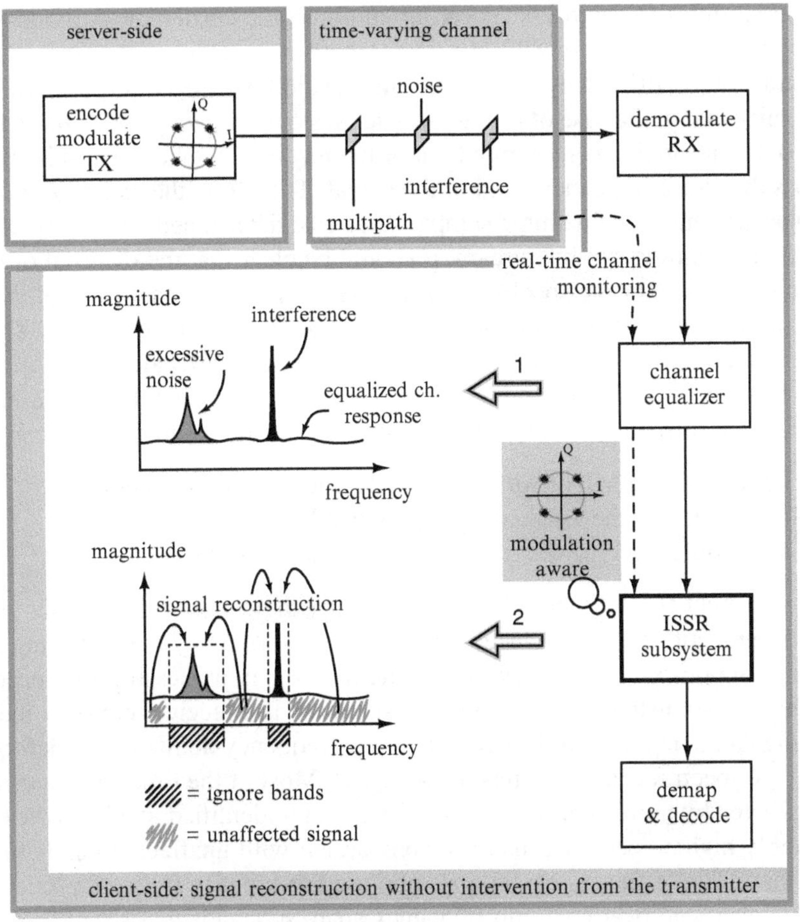

Figure 3.1. The concept of ISSR: receiver-side signal reconstruction without any intervention from server-side.

from Bluetooth transmissions make 802.11b more than often less than a pale shadow of itself. At this point, our goal is to harden 802.11b-compatible devices against such momentarily problems in the wireless connection. The one and only condition is that improvements can be implemented without breaking compatibility with the current 802.11b standard. More specifically, it is not allowed for the ISSR method to introduce changes at server-side (i.e. the transmitter), but only at client-side as the implementation details of the receiver are not imposed by the standard.

At first sight, all this could might an impossible task. At least, OFDM modulation allows the transmitter to avoid transmission over certain subbands. But the possibilities to control the spectrum of the single-carrier QPSK modulation used by the PHY layer of 802.11b do not even come close to the fine-grained spectral control offered by multicarrier modulation. However, it is important to recognize that in the case of a single-carrier system, the *energy spectral density* (ESD) of one single QPSK symbol out of the received symbol stream is spread across the entire frequency band of the signal. Therefore, the exact location of the spectral energy of a single sample cannot be determined and is dependent of the surrounding symbol stream. It should be clear that the disposal of only a limited portion of the spectral information does not necessarily result in the loss of all energy or information related to a particular QPSK symbol. Therefore, it must be theoretically possible to recover at least part of the original data from an ISI- or interferer-distorted symbol stream. For this purpose, the following information is at the disposal of the ISSR algorithm in the receiver:

- The received signal, distorted by interference or ISI due to fading
- The location of the affected frequency bands
- The modulation type (i.e. QPSK in this particular case)

The statement that it is possible to reconstruct the original signal only using the aforementioned intelligence is intuitively confirmed by the following reasoning. Suppose that the back-end of the receiver is processing the received symbol stream using blocks of n QPSK samples. The receiver converts the incoming QPSK-modulated data entities to the frequency domain, in order to inspect the spectral characteristics of the signal. Most of the time, subbands that are affected by narrowband interference are easily identified thanks to the considerably higher spectral density in comparison with unaffected bands in the spectrum. The same strategy can also be used to identify frequency-selective fading: averaged over a certain frequency span, subbands that are affected by destructive fading are very likely to display a reduced energy spectral density. After removing those parts that are affected by fading or interference from the baseband spectrum, the signal is again converted to the time domain representation. The resulting symbol stream still has sufficient SNR, but is heavily affected by ISI due to the loss of some of the non-crucial information in the frequency domain.

If enough processing power is available, it would be possible to regenerate the spectral footprint of all of the 4^n symbol combinations (Figure 3.2). Subsequently, the unaffected parts of the spectrum from the received symbol stream are compared to the spectral footprint of each of the locally generated streams. The local symbol stream which produces the best cross-correlation result with the received signal has the best chance to be the originally transmitted signal. Obviously, the enormous amount of possible combinations makes this

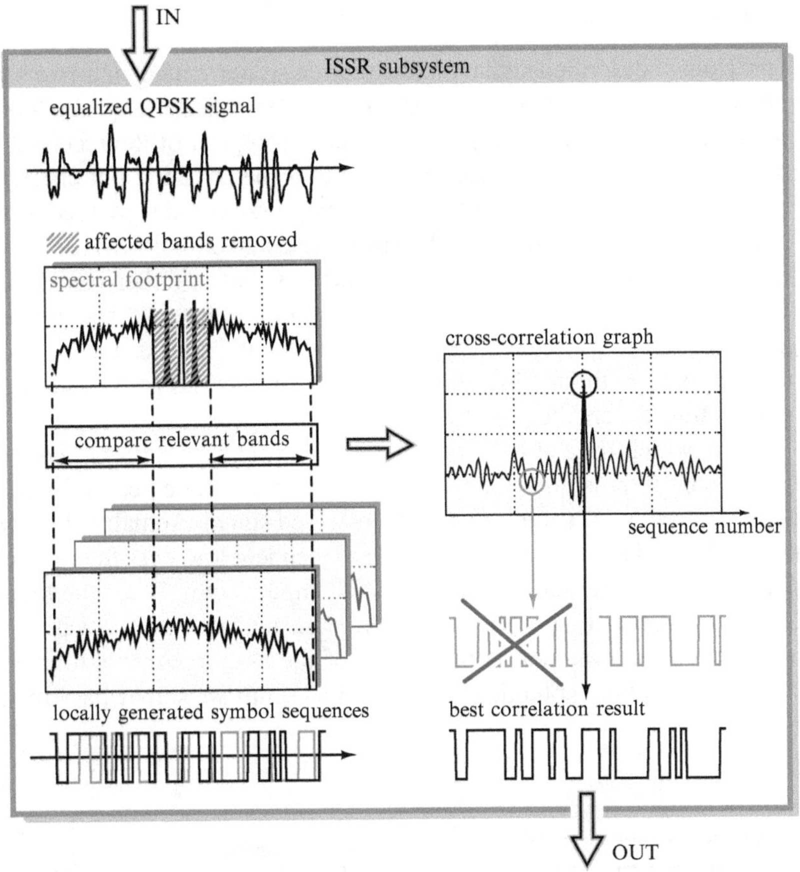

Figure 3.2. The theoretical fundamentals of the ISSR system are based on the cross-correlation of the spectral footprints of received data and locally regenerated symbol sequences.

approach – even for moderate values of n – virtually useless for any practical application. In the next section, a more efficient algorithm to retrieve the original symbol stream from the corrupted signal will be described. As a final note, it is worth mentioning that the concept of ISSR works out quite well for wideband single-carrier systems, but not for multicarrier systems such as OFDM. The underlying mechanisms of ISSR require that information is spread over a large frequency band, as they rely on the fact that frequency-selective fading will only wipe out confined parts of the spectrum. Multicarrier systems such as OFDM concentrate the signal power in smaller subbands. Since each of these subbands will experience flat fading, the prerequisites for frequency diversity are no longer fulfilled by the individual subchannels.

3.2 ISSR decoding for wideband QPSK

In the previously described signal reconstruction strategy, an exhaustive search over all possible sequences was performed in order to find the symbol combination with the best resemblance to the spectral footprint of the received data stream. Unfortunately, this approach turns out to be anything but efficient in terms of processing power. The interference suppression and signal reconstruction (ISSR) decoder technique which will be introduced below does not rely on standard equation-solving numeric algorithms, but is rather based on the working principles of a nonlinear feedback system. Figure 3.3 shows the basic block diagram of the feedback system on which the ISSR decoder is based. The forward path of the feedback loop embeds an G/s integration stage. The closed-loop transfer characteristic of this system is given by a unity gain low-pass filter of which the cut-off frequency is controlled by gain G. The input of this low-pass filter is not a single scalar value, but a static vector formed by the n consecutive QPSK samples from the received signal. Actually, the system should be regarded as a parallel system of independent low-pass filters, each of which is processing one sample from the static input vector. When the loop has achieved convergence ($[\epsilon] = \mathbf{0}$), exactly the same signal vector as was applied to the input will be available at the output of the system. At this moment, it should be clear that the potential benefits of this setup are somewhat limited.

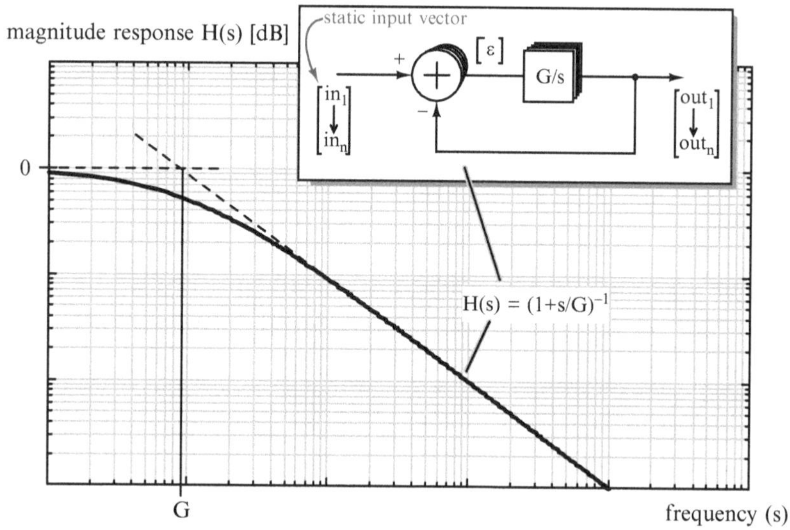

Figure 3.3. The ISSR decoder implementation is based on a low-pass filter. The feedback loop in the system tries to make the output vector equal to the input signal. The speed at which this happens depends on gain factor G.

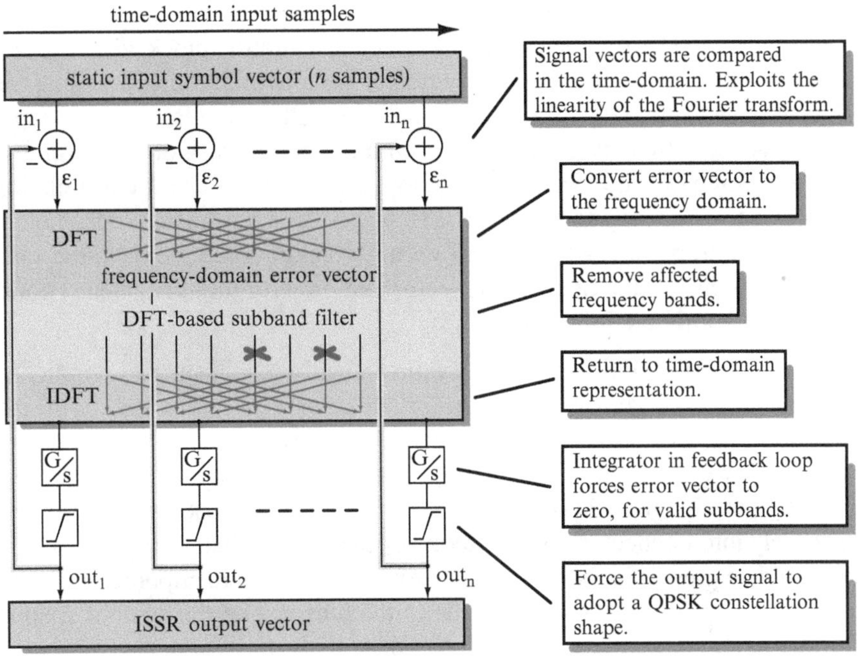

Figure 3.4. Internal functional structure of the ISSR decoder.

Remember that, as it is our purpose to reconstruct the original signal, ISI-corrupted data is available at the input. The intersymbol interference is due to the fact a part of the input data is missing: some of the subbands were zeroed out due to problems caused by destructive fading or narrowband interference. As a consequence, the error vector $[\epsilon]$ from Figure 3.3 does not contain any valid information in those particular frequency bands. To prevent that this invalid information starts poisoning the loop, the affected bands should be removed from the system before they are able to re-enter the system. In practice, this can be achieved by blocking specific frequency bands using a reconfigurable DFT filter as illustrated in Figure 3.4. It is important to note that the digital filter from Figure 3.4 operates on the error vector and uses it as a time-domain signal. In other words, the consecutive samples of $[\epsilon] = \epsilon_1, \epsilon_2, \cdots \epsilon_n$ of the error vector represent the time-domain input of the DFT filter. It is crucial not to confuse the time-domain input signal of the ISSR decoder with the time-domain dimension of the impulse response of the internal feedback system, which has a vertical signal flow in Figure 3.4.

The attentive reader may correctly remark that the current system is unable to reconstruct the ISI-free QPSK signal vector: a linear system can never produce signal components in frequency bands for which the input contains no spectral energy in the first place. For this purpose, the output of the loop is equipped with a saturation element. The complex-domain saturation block, clipping both I- and Q-signal values of the QPSK symbols, has a twofold objective. The first task is to set a limit on the maximum signal energy that circulates in the loop of the feedback system. The second and even more crucial task of this saturation block is to force the output samples to adopt the shape of a QPSK constellation. In this respect, one may consider the ISSR system as a nonlinear feedback loop of which the output is a QPSK signal vector which is forced to have the same spectral contents as the input signal vector, but only for those frequency bands that contain legitimate information.

Since there is no explicit method to mathematically model such a complicated hard-nonlinear system, the most appropriate approach to investigate the behaviour of the system under several conditions is by means of numerical simulations. However, even without digging deep into theory or simulations, it is already possible to make some preliminary predictions about the stability of the densely interconnected ISSR system: any positive feedback larger than unity that would be present in the system shall eventually result in clipping of the output vector. This clipping, which is due to the presence of the symbol-shaping saturation elements, causes spectral components in the frequency bands that are being blocked by the DFT channel filter. As a result of this, energy begins to leak out of the system which – indirectly – provides automatic gain control.

The whole idea of ISSR decoding is built around the assumption that every possible sequence of transmitted symbols has its own specific spectral footprint. Extracting the original data from the received signal vector is then only a matter of finding the most efficient way to map a particular spectral footprint to its corresponding time-domain signal vector. Obviously, every imaginable frequency-domain signal vector always has a valid corresponding time-domain representation and so has any signal which is corrupted by ISI or interference. In order to allow the receiver to reconstruct the original time-domain signal when its spectral footprint has been corrupted, a certain amount of redundant information must be included at transmission side. In a classic FEC coding scheme, this goal is commonly achieved by transforming information bits into data sequences of longer length. By allowing only a certain subset of symbol sequences, the distance between two nearby sequences of a subset can be increased. This offers the decoder the possibility to trace back to the correct information. Identical to the principles of traditional error coding, it should be possible to increase the minimum free distance between allowed signal footprints in the frequency domain.

Fortunately, the redundancy required for this purpose is automatically embedded at transmission side in the form of the modulation characteristic. The ISSR decoder for QPSK modulation takes advantage of this knowledge by gently forcing QPSK samples from the output time-domain signal vector in a QPSK constellation shape. This, in combination with a closed-loop system which continuously tries to minimize the distance between the unaffected portions of the spectral footprint of the in- and output, allows ISSR to recover the original signal. The underlying structure makes that the ISSR algorithm should be classified as a guided optimization technique, closely related to evolutionary algorithms. This finding is quite exciting because it puts things in a different light. The experimental results are delayed until after the presentation of the next section, which covers some of the practical implementation aspects of ISSR on a digital processing system.

3.3 Implementation aspects of the ISSR algorithm

The fast-varying location of deep fades in an environment with a low coherence time requires a certain level of flexibility, which is only available in a digital system. However, the feedback system introduced by Figure 3.3 is conceived as an analog, continuous-time system. In order to implement the ISSR decoder as a digital algorithm, the feedback loop must be translated from a continuous-time to a discrete-time representation. For this purpose, the continuous-time integration element G/s from Figure 3.3 must be converted from the s- to the z-domain. This is done using the *Forward Euler* (FE) transformation [Opp03] which results in the discrete-time integrator shown in Figure 3.5. The two delay elements D_1 and D_2 allow to cut the loop in two halves, which enables to implement the entire ISSR system in the form of an iterative algorithm.

Only two parameters in Figure 3.5 are still to be identified. The first gain element, $B = GT_s$, is determined both by the gain G of the continuous-time

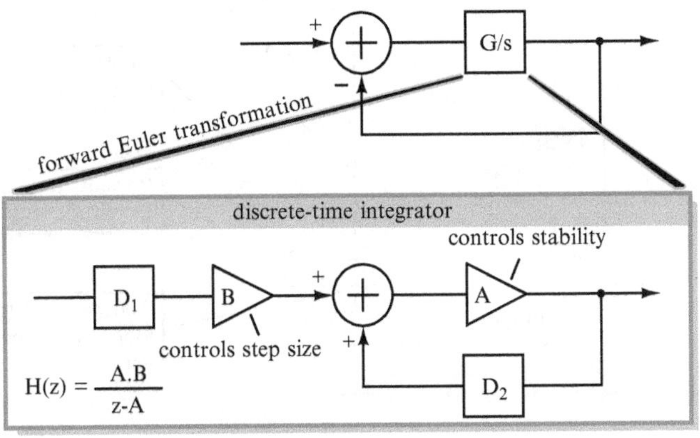

Figure 3.5. The low-pass filter from Figure 3.3 can be converted to the z-domain using the forward Euler transformation.

integrator and by the sample period T_s of the discrete-time system. Once again, note that this time interval is not related in any way to the symbol period of the receiver. Parameter T_s has no associated physical meaning, but is related to the step size of the algorithm. Reducing this step size will increase the accuracy of the ISSR algorithm, but requires a higher number of iterations before convergence will be reached. Finding a closed-form expression for the relationship between accuracy, convergence time and gain B is not straightforward due to the nonlinear nature of the loop. The same argument holds for gain element A, so acceptable values for these parameters should be derived using numerical simulations. A good initial value for both B and A was found to be very close to unity.

Remark that when factor A is chosen equal to 1.00, the pole of the integrator is located on the z-plane unity circle. The result is a marginally stable system with a high risk to unbounded amplitude levels inside the loop. This can be easily prevented by a restriction of the internal signal swing in the discrete-time integrator. When considering the topology from Figure 3.4, this only requires a relocation of the saturation blocks from the output of the ISSR loop to the forward path of the discrete-time integrator. The final result, after a slight repositioning of the delay elements, is shown in Figure 3.6. At this moment,

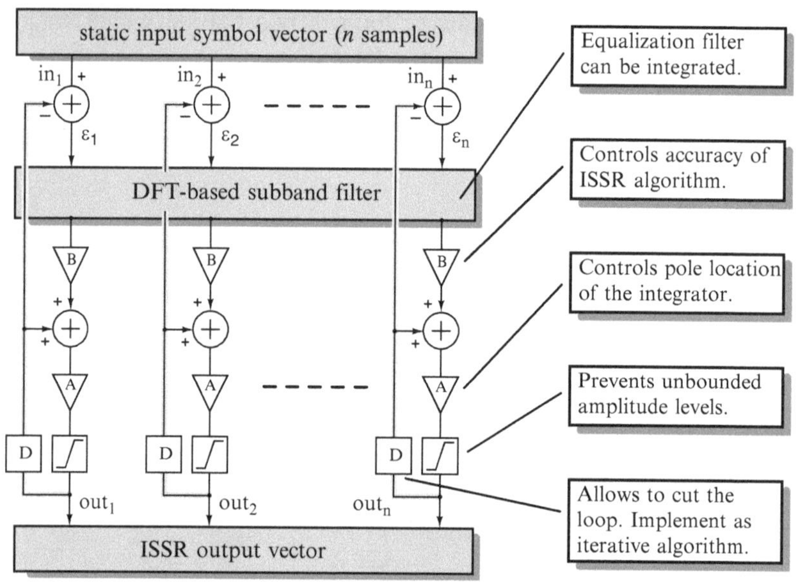

Figure 3.6. Discrete-time implementation of the ISSR loop. Some smart tuning of the gain factors A and B can considerably speed up the convergence process of the algorithm.

it is even allowed to place the pole of the integrator outside the unity circle. In fact, the potentially unstable nature of the system when the value of A is chosen slightly larger than unity has a beneficial effect: the output of the ISSR loop will tend to float and clip against one of the boundaries imposed by the saturation blocks, thereby enforcing a QPSK constellation shape of the symbols in the output vector. Numerical simulations have shown that this leads to a significant improvement in the convergence speed of the decoder.

The discrete-time implementation of the ISSR loop in Figure 3.6 also allows for some inventive extensions: the DFT-based subband filter, which prevents that corrupted parts of the spectrum can gain control over the signal reconstruction process, could be fused together with the role of the channel equalization filter. Examination of the spectral contents of the error vector $[\epsilon]$ in Figure 3.6 can provide useful information for this purpose. Another improvement to ISSR is to employ an adaptive step size for the algorithm. Once more, this could be easily achieved by monitoring the ESD of the error vector. At the start of the iteration cycle, choosing a larger step size can considerably boost the convergence speed. Towards the end of the convergence process, the overall accuracy is then improved by decreasing the sample interval T_s. Remember that this can be done by a simply reduction of the gain factor B in the discrete-time topology.

A final suggestion that could improve the robustness of ISSR in a frequency-selective fading channel would be to add *pilot symbols* at transmission side, comparable to the pilot tones used in OFDM. Pilot symbols are predefined training symbols at specific locations in the transmitted data stream, known by both the transmitter and the receiver. The output vector of the ISSR decoder can be forced to adopt these pilot symbols, which improves the robustness of ISSR under adverse channel conditions.

3.4 Performance of the ISSR algorithm

A reduction of the available (i.e. unaffected) frequency spectrum in a wideband communication system due to narrowband interference or frequency-selective fading leads to a significant bit error rate (BER) penalty in case of an uncoded transmission. Let F be the fraction of the available bandwidth and 1-F the fraction of the transmission channel that is corrupted by interference or some other dark force. Also consider the ideal case where the available fraction is completely free of noise. The elimination of the corrupted fraction by zeroing out the affected subbands can be modeled by colored Gaussian noise. As a consequence, the performance of an uncoded transmission system (without signal reconstruction) can be approximated by the following complementary error function (3.1):

$$\begin{aligned} \text{BER} &= Q\left(\sqrt{\frac{\text{S}}{\text{N}_{\text{color}}}}\right) = Q\left(\sqrt{\frac{F}{1-F}}\right), \quad (3.1) \\ \text{where} \quad Q(\text{z}) &= \frac{1}{2}\,\text{erfc}\left(\frac{\text{z}}{\sqrt{2}}\right) \\ \text{and} \quad \text{erfc}(\text{x}) &\triangleq \frac{2}{\sqrt{\pi}}\int_{\text{x}}^{\infty} e^{-\lambda^2}\,\text{d}\lambda \end{aligned}$$

According to Shannon's limit [Cou97, Ben02], the ideal coded communication system requires at least that $E_b/N_0 \geq -1.59$ dB for error-free communication. For a QPSK system, this corresponds to a minimum required uncorrupted fraction $F = 58\%$ of the total channel bandwidth.[1] Today's best high-performance error correcting codes, such as *Turbo codes* [Ber93], can provide a BER as low as $1/10^5$ at an E_b/N_0 of +0.7 dB. This would correspond to an interference-free bandwidth of at least $F = 70\%$. The performance of the ISSR decoder was evaluated by simulating the bit error rate as a function of the available bandwidth F. The simulation was performed using a transmission block length of $4k$ symbols, while the gain factors A and B from Figure 3.6 were set to $A = 1.05$ and $B = 1.00$, respectively. In order to obtain maximum performance, 30 iterations of the ISSR loop were performed before the BER at the output of the ISSR decoder was computed. Figure 3.7 shows the bit error rate characteristic of the ISSR decoder as a function of the bandwidth fraction F and the normalized signal to colored noise ratio E_b/N_c. The upper bound for the coding gain of the ISSR decoder with respect to the uncoded reference transmission system is approximated by Equation (3.2):

$$G_{\text{ISSR}} = -10\log_{10}\left[1 - F\log_2\left(1 + \frac{F}{1-F}\right)\right] \quad (3.2)$$

On the performance plot of Figure 3.7, one can distinguish two main operating regions. Turbo coding is clearly the most performing error correction mechanism above the $E_b/N_c = 0.3$ dB limit. For the region below 0.3 dB however, the ISSR decoder provides better performance. The main reason for this is that the ISSR algorithm is supported by additional information such as the modulation type and the location of problematic frequency bands. Simulations have demonstrated that the ISSR decoder can be concatenated to the Turbo coder.

[1] Using $\text{SNR} = \eta \cdot E_b/N_0$. Bandwidth efficiency η is 2 bits/s/Hz for QPSK using raised cosine filtering with a rolloff factor $r = 0.0$ ([Cou97] p. 351, 575).

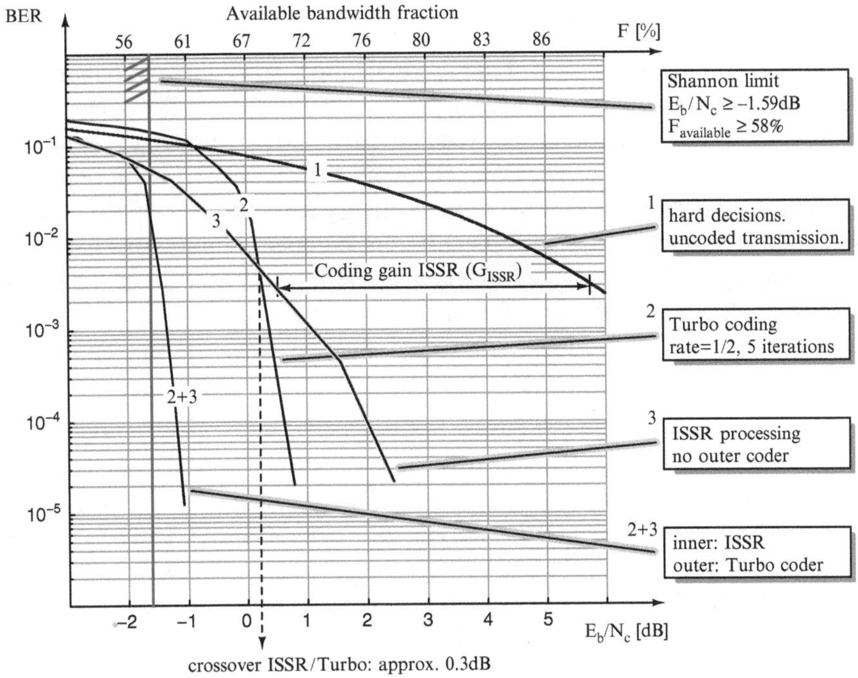

Figure 3.7. Simulated performance of ISSR as a function of the available bandwidth. When ISSR is the inner coder of a Turbo coder, the overall performance is within 0.4 dB of the Shannon limit.

Using the ISSR algorithm as the inner decoder, the performance of the compound system comes within 0.4 dB of Shannon's theoretical limit. It is important to realize that, while coding techniques merely based on redundancy cause a reduction in the effective throughput, the ISSR decoder does not affect the data rate of the system.

The throughput of a wireless link is one way to measure its performance, however that's not the whole picture. In a practical implementation with limited energy resources, it is important to properly distribute the available processing power between ISSR and the outer coding mechanism. For example, in a channel that is uniformly contaminated by AWGN noise, it does not make sense to waste energy in the ISSR decoder. Also, Turbo coding costs around 10 times more computing power than a viterbi FEC decoder [Des03]. In a severe multipath environment, it is certainly worth the consideration to concatenate ISSR with a less complex FEC coding scheme. The reduction in power consumption will more than compensate for the implementation loss that comes with a less sophisticated error coding scheme.

3.5 ISSR under non-ideal circumstances

The simulation results presented in Figure 3.7 did assume that unbounded processing power is available to the receiver. The limited availability of computing resources will force a trade-off between bit error rate performance, latency and throughput efficiency. For example, reducing the number of iterations in the ISSR algorithm could provide significant energy savings in a battery powered system. The BER characteristics for different numbers of iterations have been plotted in Figure 3.8. Remark that most of the processing gain of ISSR is already achieved during the first few iterations. The final number of iterations depends on the price one is willing to pay to keep down the implementation losses (IL).

For example, suppose that our target BER is $1/10^3$. Only four iterations are required to achieve a coding gain of 2 dB and to cross the $E_b/N_c = 5$ dB boundary. A further reduction of the IL by another 2 dB is achieved by augmenting the number of iterations to 12. This is a viable option, also in a battery powered device. As was mentioned earlier, increasing the step size of the ISSR algorithm speeds up the convergence process at the cost of the accuracy. This time though, there is no penalty for slightly increasing the step size: the limited

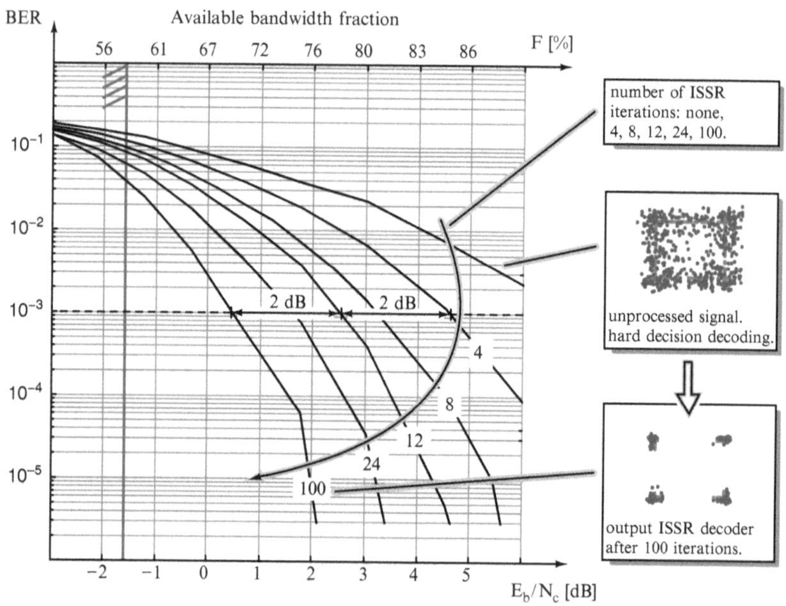

Figure 3.8. Simulated performance of the ISSR decoder as a function of the number of internal iterations. Most of the processing gain is achieved during the first few iterations.

number of iterations will reduce the accuracy anyway. In the example from above, increasing the step size by approximately 50% ($B = 1.50$) allowed to reduce the number of iterations from 12 down to 8 without noticeable impact on the predetermined bit error rate performance (BER $\leq 1/10^3$ for $E_b/N_c = 3$ dB).

Also shown in Figure 3.8 is a visualization of the QPSK symbol constellation at the in- and output of the ISSR decoder. Before signal reconstruction, the constellation points are spread over the entire signal plane as a result of the effects of ISI. During the reconstruction process, the data points slowly start to cluster around one of the four QPSK constellation points. It is very intriguing to observe how some of the sample points migrate to their final position, as it is not necessarily the shortest path that is followed between their start and end position. On the contrary, some sample points tend to move towards one constellation point, then seemingly for no apparent reason 'decide' to switch direction, in order to end up in one of the other three quadrants.

This behaviour is comparable to the behaviour of individuals observed in *particle swarm optimization* (PSO) algorithms. At the start of the optimization process, the particles have a large *cognitive* component which forces them to propagate individually in the direction of the nearest constellation point (they don't care about their fellow team members). After a few iterations of the ISSR algorithm, most particles approach their optimum position. At this moment, the residual energy in the error vector becomes dominated by those few particles that were initially forced towards a wrong constellation point. Due to the effects of the integrator in feedback configuration, increasingly opposing forces start to build up. This is where the *social* component of PSO comes into play. Most individuals in the swarm stay at their current position, since their composite spectral signature resembles the required spectral footprint quite well. Thanks to the joint forces of individuals in the swarm, only the few remaining particles (QPSK symbols) with an adverse contribution to the spectral match are driven to a new location in the constellation diagram. This despite the individual, cognitive component which initially forced them towards a wrong constellation point.

Performance of ISSR in noisy channels

All previous simulations have been performed under the assumption that the available fraction F of the transmission channel is completely free of additive white Gaussian noise (AWGN). In reality, the overall signal quality at the input of the ISSR decoder is determined not only by the colored noise caused by the elimination of ISI or interference, but also by the thermal noise floor of the channel. Define E_b/N_0 as the bit energy per AWGN power spectral density and

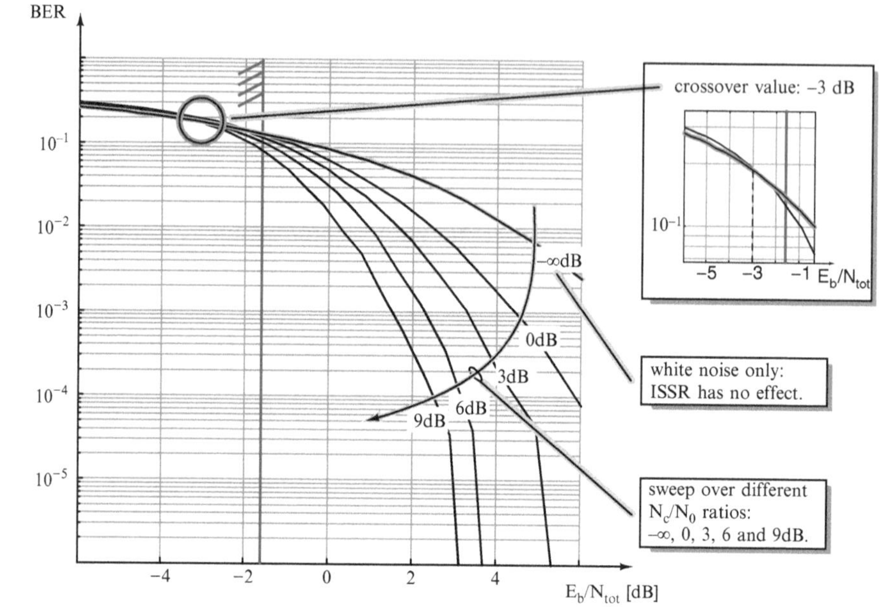

Figure 3.9. Simulated performance of the ISSR decoder for different colored noise (N_c) versus white noise (N_0) ratios. If only white noise is present in the signal, ISSR offers no advantages.

N_c/N_0 as the colored noise versus white noise ratio. The overall signal quality E_b/N_{tot} is then expressed as (3.3):

$$\frac{E_b}{N_{tot}} = \frac{E_b/N_0}{1 + N_c/N_0} \tag{3.3}$$

The BER characteristics for various N_c/N_0 values are plotted in Figure 3.9. As might be expected, the performance of the ISSR decoder drops when the noise floor of the channel is being increased. But what is of more importance is the location of the crossover point between the curves of ISSR decoding and hard-decision demapping (topmost curve in Figure 3.9). Most error coding schemes exhibit some sort of *coding threshold* below which the performance of the decoder actually becomes worse than the error rate of the uncoded system. It turns out that, for all simulated ratios of N_c/N_0, the crossover point of the ISSR decoder is located at $E_b/N_{tot} = -3$ dB. Interestingly enough, this is exactly the point where the SNR of the QPSK signal intersects the 0 dB boundary.[2] Below this value, the amount of noise energy exceeds the energy of useful information as a result of which the ISSR decoder tries to correct the input signal

[2] $\text{SNR} = E_b/N_0 \cdot \eta_{max}$, where $\eta_{max} = 2$ is the maximum achievable bandwidth efficiency for QPSK modulation using a raised cosine-rolloff filter with rolloff factor $r = 0.0$ ([Cou97] p. 351).

based on mostly erroneous information. This remarkable observation leads to the assumption that the ISSR decoder does not have a deteriorating effect on the signal quality for any practical E_b/N_{tot} value. In this sense, the decoder does not inject additional noise that could compromise the performance of subsequent decoding stages.

At the start of the ISSR algorithm, it is important to initialize the output vector of the ISSR algorithm to the received input vector. When doing this, simulations show a fairly stable crossover value around -3 dB, even for a low number of iterations in the ISSR routine. By avoiding a randomly chosen initial population, the optimization loop converges much faster to the final steady state value, which is of special interest for applications with a limited processing power. Also note that, unlike most PSO implementations, the ISSR routine does not employ probabilistic mutations during the optimization process. This ensures a smooth migration from the initial input signal to the reconstructed signal vector. In most other optimization contexts, mutations are introduced to prevent that the algorithm gets trapped into local optima. However, initializing the output vector of the ISSR algorithm to the input signal appears to result in a fairly smooth optimization surface. Also, the effect of relentless grinding away useless frequency bands is smeared out over a wide range of symbols in the time domain. This can be attributed to the repeated conversions between the time and the frequency domain representation of the signal. This results in relatively smooth transitions between the consecutive QPSK vectors and a fairly progressive increase of signal quality at the output of the ISSR subsystem.

A final observation from Figure 3.9 is that, when the contribution of AWGN noise to the total noise figure increases, the BER performance curve of the ISSR decoder reverts to the scores of hard-decision QPSK demapping. This finding is consistent with the internal workings of the signal reconstruction algorithm: ISSR offers no protection whatsoever against the unpredictable nature of white noise in the time domain: the information contained in a single QPSK symbol sample cannot be retrieved from surrounding symbols as ISSR is focussed on the exploitation of frequency diversity, instead of diversity in time. This is also the reason why the performance of ISSR does not show the characteristic steep slope as typically seen in sequence-based coding schemes. In order to fully benefit from its advantages, ISSR should thus always be used as the inner decoder of a compound error correction system. The result is a robust coding system that exploits diversity in both the time and the frequency dimension.

If correctly dealt with, frequency diversity and intersymbol interference can actually become one of the major advantages of a wideband, frequency-selective channel: a narrowband communication system will always suffer from flat fading in a multipath channel. Increasing the transmission power in a narrowband channel does not significantly increase its performance because of the proportional increase of power in the multipath reflections. In a fast fading channel this is not necessarily a problem: repetitive short outages of the communication link are easily overcome with an appropriate FEC mechanism. However, the duration of deep fades becomes substantially longer when both the narrowband transmitter and receiver are stationary devices in a quasi-stationary environment. Under these circumstances, there is very little forward error correcting codes can do to survive.

Chapter 4

BENEFITS OF ISI IN THE INDOOR ENVIRONMENT

In this section, the problems caused by a multipath channel are discussed in more detail. First, a short descriptive overview will be given on the specific categories under which a multipath environment can be classified. Along with this, it is indicated how this is anticipated for in some of the wireless systems (e.g. 802.11a/b/g) that have been discussed earlier in this book. Towards the end of this section, it should become clear to the reader under which particular circumstances the extra costs of using a special architecture such as the multi-finger RAKE receiver or a multi-antenna setup can be justified as a way to increase diversity and as such also the reliability of a wireless link. First, in the next few sections, some important concepts such as the delay spread of the radio channel will be introduced. After this, the impact of the properties of the wireless channel on the reliability of the link are discussed. It will be pointed out that, in contrast to common belief, intersymbol interference is actually a necessary precondition to improve the reliability of a wireless system that is operated in a multipath indoor environment.

4.1 Power delay spread

The problems caused by multipath and fading – the latter one being essentially the frequency domain representation of the equivalent multipath channel response – have been mentioned several times before. However, a superficial use of the word 'multipath' ignores the fact that this term covers a very broad range of possible channel conditions. The actual properties of the wireless channel as experienced by the user depend on several parameters such as the spectral bandwidth being used and the stability of the channel over a certain period of time. Probably the most important property is the time frame during which delayed versions of the same transmitted symbol arrive at the antenna of the

W. Vereecken and M. Steyaert, *Ultra-Wideband Pulse-based Radio: Reliable Communication over a Wideband Channel*, Analog Circuits and Signal Processing Series,

receiver. The extent in which a single impulse applied to the input of the channel is spread over time at the output is described by the *delay spread* parameter of the channel. Rather than describing the delay spread as the time interval between the first and the very latest arriving signal component, it is more convenient to use the RMS delay spread. Formally, the RMS delay spread (σ_τ, τ_{RMS}) is defined as 'the square root of the second central moment of the power delay profile' [Hol99]. The power delay profile (PDP), in his turn, is a statistical distribution which describes the average expected power that arrives at the antenna as a function of the excess time delay, starting from the moment the impulse is injected into the channel by the transmitter.

Let us explain this more carefully using Figure 4.1. The time-varying impulse response $h(\tau, t_1)$ of the wireless link is defined as the response of the channel at time τ (which is the propagation delay from transmitter to receiver) to a single impulse applied at time stamp t_1. The mathematical expression for the power delay profile is then given by the average power at the receiving antenna as a function of the propagation delay τ (4.1):

$$\text{PDP}(\tau) \quad = \quad \text{E}\left[|h(\tau, t)|^2\right] , \tag{4.1}$$

where E[·] is the expectation operator.

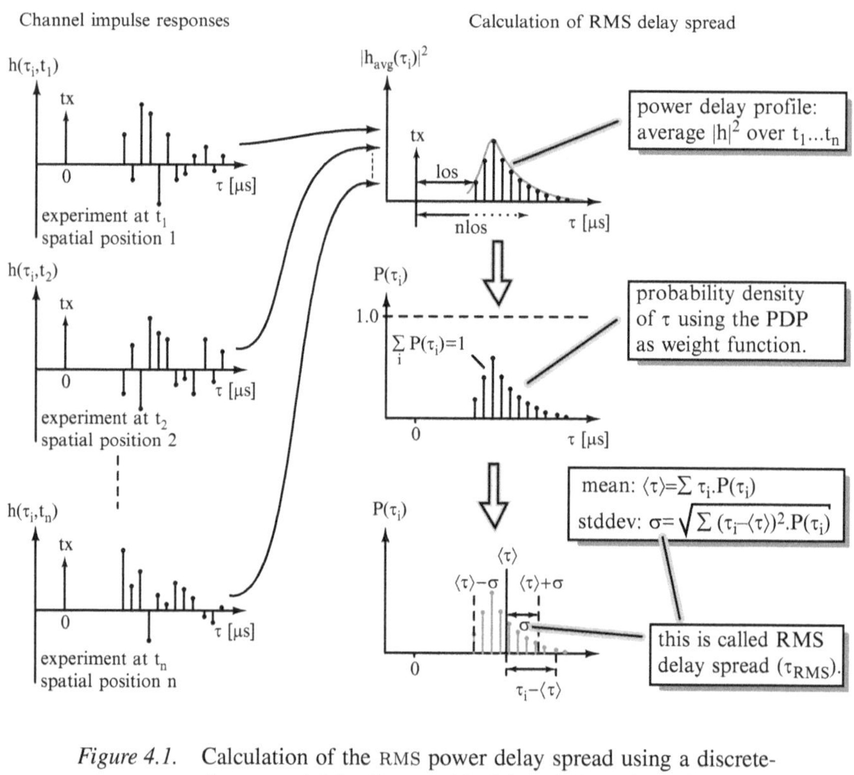

Figure 4.1. Calculation of the RMS power delay spread using a discrete-time tapped delay line model of the wireless channel.

In more intuitive terms, the power delay profile of a channel gives the average power at its output for a given propagation delay τ. Obviously, each delay time τ is always related to a certain physical propagation path length in the multipath channel. The shape of the power delay profile is thus determined by the type of the channel.

For example, the PDP of a heavy multipath indoor channel is typically modelled by an exponentially decaying function [Rud06]. This is mainly due to the length of the different paths that the signal follows from the transmission antenna to the receiver. Signals that have a longer propagation time τ not only have travelled a longer distance, but in the indoor environment, they also have undergone more reflections than the first arriving signal component. As a consequence, the power delay profile of an indoor channel will show that, on average, multipath components with a longer propagation time will also have a higher path loss (PL) coefficient.

If the distance between transmitter and receiver is increased, such as is the case in an urban environment, the impulse response of the channel is spread over a longer time period. The delay profile of such a channel contains multiple individual clusters of components. It is not always true that the cluster with the shortest propagation delay will also have the highest power, but the power decay of individual signal components within each cluster is still exponential [Bur98]. This behaviour is explained by the observation that, apart from local scattering which causes the exponential PDP within a cluster, large remote objects (e.g. a building) can cause strong reflections with a high propagation delay. Probably the best example of a system with an urban multipath delay profile is encountered in the cellular communication system (GSM). The RMS delay spreads found in an urban environment are in the order of 1 µs, equivalent to a difference of more than 300 m in the length between two individual propagation paths [Sou94].

It should be clear by now that merely the time spread between the earliest and the latest component is not a good measure of how the channel spreads its energy over time. For this reason, the RMS delay spread of the channel is used to indicate how energy is being spread. The RMS delay spread is defined as follows. First, the probability density function (PDF) of the power delay profile is calculated (4.2):

$$f_{\mathrm{PDP}}(\tau) = \frac{\mathrm{PDP}(\tau)}{\displaystyle\int_0^\infty \mathrm{PDP}(\tau)\,d\tau} \tag{4.2}$$

Since the impulse response of the channel is a causal function, the value of the PDF is zero for $\tau < 0$. The RMS delay spread of the channel is then obtained by evaluating the square root of the second central moment of the probability density function of the power delay profile [Dev87]. In more familiar and comprehensive terms, the RMS delay spread is given by the standard deviation of $f_{\text{PDP}}(\tau)$, as shown in (4.2):

$$\tau_{\text{RMS}} = \left[\int_0^\infty (\tau - \langle\tau\rangle)^2 \cdot f_{\text{PDP}}(\tau)\, d\tau \right]^{\frac{1}{2}} \quad \text{[unit: s]}$$

$$\text{with} \quad \langle\tau\rangle = \int_0^\infty \tau \cdot f_{\text{PDP}}(\tau)\, d\tau \qquad (4.3)$$

In practice, the RMS delay spread is calculated from statistical data obtained by real-life measurements of the delay profile of the channel (Figure 4.1). Some typical measurement results for different environments are shown in Table 4.1 [Skl01].

Environment	Frequency	RMS delay spread	Notes
Urban	910 MHz	avg. 1.3 µs	NY City [Cox75]
Suburban	910 MHz	0.2–0.3 µs	Typical avg. [Cox72]
Indoor	850 MHz	<0.27 µs	Office building [Dev90]

Table 4.1. Typical values of delay spread in the 900 MHz band.

A very important side note here is that the power delay profile can indeed be used as a measure for the *average* combined signal power available from all multipath components. However, as a result of destructive interference, it will not always be possible for a receiver to extract all of the available signal power from the channel at any moment in time. This statement is valid for any single-antenna receiver setup, no matter how complex the receiver architecture may be. A well-known example of this is the fading experienced by the narrowband broadcast FM radio service.[1] Interference between two multipath components of the same signal will cause considerable variations in the available signal power. For a mobile FM receiver, such as a car radio, the listener will experience fading as periodic audio drop-outs, caused by the squelch[2] circuit of the receiver. In a stationary receiver, simply moving the antenna over a small distance can solve the problem most of the time. Fortunately, fading very rarely poses a real problem in the FM broadcast service: broadcast FM features a sufficiently high transmission power (up to more than 100 kW[3]), as a result of which the average RF signal power at the client side is at least an order of magnitude higher than the sensitivity level of the receiver. As a result of this,

[1] The term 'narrowband' refers to the 150 kHz wide channel of a commercial FM broadcast station. It should not be confused with the narrowband FM (N-FM) modulation method.

[2] A 'carrier squelch' or 'noise squelch' mutes the audio output in case of an insufficient signal strength.

[3] One nasty (but amusing) side-effect: the excessive transmission power of a nearby radio station can desensitize the RF front-end of a receiver over its entire frequency band [Han67, Gav84a, Gav84b].

broadcast FM nearly always survives serious drops in the signal quality without noticeable interruptions. Somehow, it may not surprise the attentive reader that boosting the transmission power a few orders of magnitude may not entirely be the most energy efficient solution for a battery-powered communication system. It is the goal of this section to find better ways to deal with the unstable character of the multipath channel, without losing out too much in terms of performance and power efficiency.

4.2 Frequency-selective versus flat fading

The power delay spread is a very important statistical parameter during the design of a wireless communication system: it provides a measure for the time dispersion of the signal as seen at the antenna terminal of the receiver. For every symbol that the transmitter injects into a multipath channel, the receiver will pick up several delayed versions of that symbol, each with a randomly distributed attenuation and phase rotation. If the duration of a symbol (T_S) becomes shorter than $10 \cdot \tau_{RMS}$, then delayed versions of one transmitted symbol start to interfere with adjacent (future) symbols. In the time domain this phenomenon is referred to as the well-known intersymbol interference (ISI) problem. From the equivalent frequency domain point of view, this type of wireless channel is characterized as *frequency-selective*.

On the other hand, when the duration of a single symbol becomes longer than a few times the delay spread, the consecutive symbols will not corrupt each other, but the interference now takes place within the duration of a single symbol (Figure 4.2). From a frequency-domain point of view, the difference in the length of the propagation paths causes a random phase rotation. As a consequence, individual signal components that arrive at the receiving antenna within one symbol period can interfere in a constructive or a destructive way with each other. Within the frequency band of a narrowband receiver, this effect is experienced as *flat fading*: all frequencies in the whole frequency range of the wireless link experience the same path loss. Destructive interference is a major issue under flat fading circumstances, because the received signal can completely disappear at any moment, no matter how much RF signal power is being supplied from transmission side.

Apart from the example of broadcast FM, another nice illustration of a frequency-flat fading channel is provided by the periodic fading of shortwave radio signals. Due to the partial reflection of radio signals at different heights in the ionospheric layers of the earth, several reflected versions of the same signal arrive at the receiving antenna with a different delay. The very typical periodic fading of shortwave radio stations is caused by the vertical movements of the ionospheric layers, which results in an alternating sequence of periods with constructive and destructive interference.

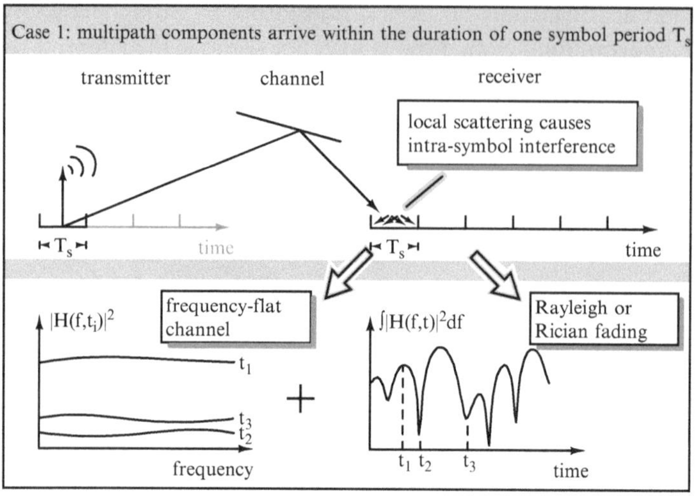

Figure 4.2. A transceiver system experiences frequency-flat fading if the duration of the transmitted symbols is longer than the delay spread of the channel.

The loss of signal that is caused by frequency flat fading is a problem which is very difficult to avoid. In a digital transmission system, it may be possible to survive a temporary loss of the signal by employing an error coding mechanism which offers a sufficient level of time-diversity. However, this approach is known to not perform very well in a quasi-static environment. If both the transmitter and the receiver are stationary or moving very slowly, the duration of a deep fade interval can become longer than the maximum time frame over which the error coder is able to spread its data needed to reconstruct the original information. Moreover, in a real-time communication system, the use of coding schemes with a better performing time-diversity is not always allowed. This is, for example, the case in low-latency communication systems such as GSM. To avoid long-duration fades, the GSM system employs a slow frequency hopping mechanism. During the transmission period of a single transmission slot (0.577 ms), the frequency remains unchanged. Different packets however, are transmitted at different frequencies at a rate of 217 hops/s.[4] Even when operated under static channel conditions, it seems that the addition of the artificial extra layer of frequency diversity can preserve the communication link from unacceptable long outages. By now, it should become clear that a flat fading channel offers less opportunities to counter the effects of interference than a frequency-selective fading channel does. In the latter case, of course,

[4]The hopping algorithm is distributed by the base station over the broadcast control channel (BCC).

intersymbol interference will seriously cripple the performance of the link if an unsuitable modulation method or a totally wrong coding strategy are thrown into the game. However, if an efficient way can be found to disentangle the different delayed signal streams at the receiving antenna, the problem which causes ISI can be turned into a surprising strength of the frequency-selective multipath channel.

DEFINITION – COHERENCE BANDWIDTH OF A CHANNEL

The coherence bandwidth (B_c) is a measurement of the average frequency span over which the frequency response $H(f)$ of a channel can be considered as approximately 'flat' and it is not just a coincidence that it is defined using the RMS delay spread of the channel (4.4):

$$B_c = \frac{1}{2\pi \tau_{RMS}} \quad \text{[units: Hz]} \tag{4.4}$$

Note: the interpretation of coherence bandwidth given here is only very basic. For a more rigorous treatment, see [Jak94] and [Bar05].

An easily overlooked aspect of frequency-selective fading channels is that, in order to effectively be able to make advantage of their *frequency diversity*, the total bandwidth covered by all subbands together must be sufficiently larger than the coherence bandwidth (see inset 4.4) of the channel. If this is not the case, the wireless link can be easily corrupted since the entire frequency-selective channel can suddenly disappear in a deep fade. The problem here is not caused by ISI, but by destructive interference caused by the multipath components that arrive within the interval of a single symbol period (Figure 4.3).

This form of destructive interference within a single symbol interval, called *intra-symbol interference*, causes frequency-flat fading and comes on top of the frequency-selectivity of the channel. In other words, the received signal power will not only change over frequency, but the total combined power over all subbands will also suffer from fading over time, which results in intermittent link outages. A transmission with a sufficiently wide spectral footprint with respect to the coherence bandwidth of the channel has less chance that all frequency bands fade away at the same moment (Figure 4.4). It follows that ISI is indeed not only an annoying side-effect, but also a necessary precondition for a frequency diverse – and thus reliable – wireless link.[5]

[5] The underlying problem described here is in fact related to the *multipath resolvability* of the system, and will be discussed in more detail in Section 4.4.

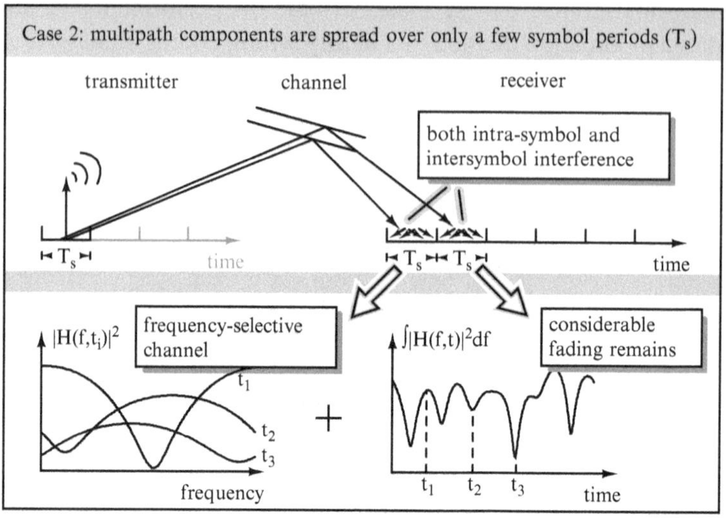

Figure 4.3. If the channel response is only slightly longer than the symbol period, this results in a frequency-selective channel. However, the received power still has a considerable level of fading.

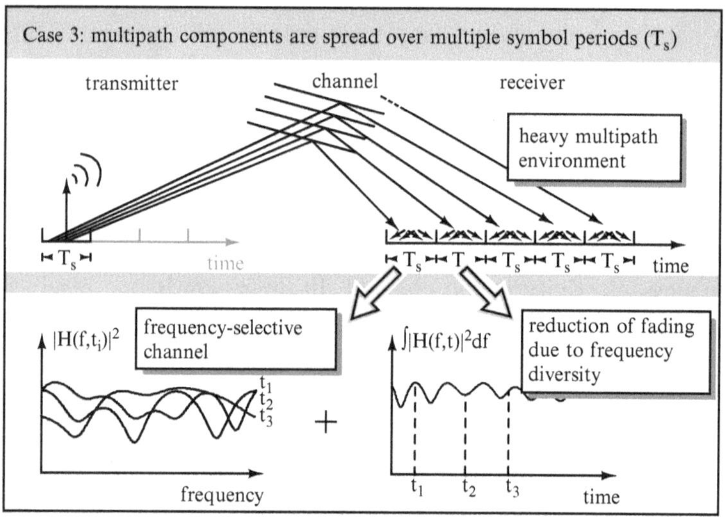

Figure 4.4. A heavy multipath environment causes a lot of intersymbol interference, but also results in a more reliable channel as a result of frequency diversity.

Case study: frequency diversity in the OFDM system

It was shown earlier that OFDM solves the problem of ISI by splitting up the data stream over *n* parallel narrowband transmission channels. If the symbol duration $T_{s,n}$ applied to subchannel *n* is longer than the delay spread of the channel, then ISI is avoided and – in theory – will not cause any additional signal losses. The multipath resolution capabilities of OFDM lie in the fact that it is possible – at least in theory – to redistribute the data load from subchannels with an insufficient signal quality to channels with a better signal-to-noise ratio. No signal energy is lost or created due to ISI in a frequency-selective channel, so the Shannon capacity is not affected under ideal circumstances.

Of course, this is only true when the appropriate subband bit-loading techniques are employed. Failing to do so indeed leads to performance loss, but this effect should be classified as a problem of implementation loss (IL) rather than a fundamental limitation of OFDM. Since adaptive bit-loading requires a difficult-to-implement support from transmission side, it is currently not supported by the 802.11a/g standards. All subchannels share the same modulation depth, which has a serious impact on the performance of the 802.11a/g system. Where in the ideal case of adaptive bit-loading the loss of SNR due to destructive intersymbol interference in one band is regained in another subband, failing to support adaptive loading downgrades the performance of OFDM: channels with constructive interference are not used at their peak capabilities, while the modulation depth is too high for subchannels with destructive fading.

For these reasons, bit-interleaved coded modulation (BICM) was introduced in the 802.11a/g standard [Cai98, Set06]. Interleaving information across the OFDM subchannels merely spreads the excessive BER experienced in some of the subchannels uniformly over a larger sequence of data bits. Bit-interleaving in a non-adaptive OFDM scheme must be seen as a method which transforms the problem of ISI into a problem of white noise, an issue which can be easily solved by standard (modulation-unaware) error coding schemes. In this respect, the way in which the 802.11a/g system employs OFDM has very little to do with true RAKE receiver architectures, in which it is attempted to resolve and recombine as much of the energy of several multipath components as possible.

4.3 Coherence time

Apart from a mobile transmitter–receiver setup, the varying characteristics of a wireless channel are also caused by moving objects in the propagation path from the transmitter to the receiver. In either of the foregoing cases, the net result is a changing length of the propagation path over time. From the point of

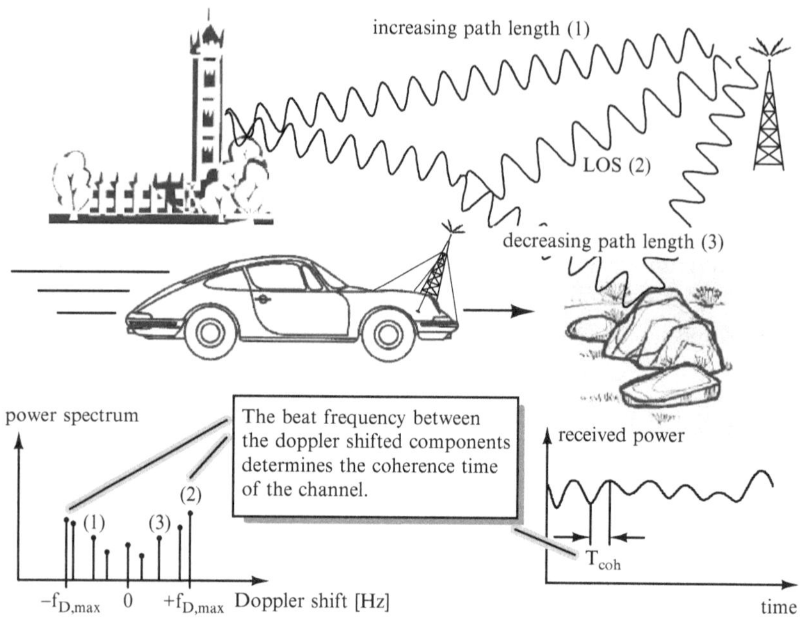

Figure 4.5. The beat frequency (BF) between the extreme Doppler-shifted frequencies causes an alternating sequence of positive and destructive fades every half wavelength of the BF. This determines the coherence time of the channel.

view of the receiver, this process is perceived as a change in the wavelength or a shift in the carrier frequency of the transmitter. This phenomenon is known as the *Doppler effect*. Of course, in the indoor channel, there are multiple propagation paths from transmitter to receiver, each of them with a different *Doppler shift*. The maximum deviation from the original center frequency is called the *Doppler spread* ($f_{D,\,max}$) of the signal. If there is a direct path between transmitter and receiver, the Doppler spread can be obtained directly from the relative velocity (v_{rel}) between both devices. The maximum Doppler shift is then given by $f_{D,\,max} = v_{rel}/\lambda$. For an indirect NLOS propagation path, however, the speed at which the length of the path changes does not only depend on the incident angle of the multipath component on the reflecting surface, but also on the velocity of the object itself (Figure 4.5).

Typical values for the Doppler frequency shift start at a few Hertz and are often limited to a few hundreds of Hertz, so the frequency offset itself won't be too much of a headache for most applications. Rather, the underlying cause which is described by the Doppler spread forms the real problem: the arrival time of the multipath components changes over time, and so do the amplitude and the phase of the channel response. The higher the Doppler spread of the channel is, the faster the characteristics of the channel are going to change over time and the more difficult it is for a wireless system to keep track of the channel response. In fact, it becomes very difficult for the receiver to discern between intentional phase changes caused by modulation of the signal-of-interest and the modulation of a multipath component caused by the channel itself.

DEFINITION – COHERENCE TIME OF A CHANNEL

The stability of a channel over time is characterized using the coherence time T_c, which is defined as the inverse of the maximum Doppler spread frequency (4.5):

$$T_c = \frac{1}{f_{\text{Doppler, max}}} \quad \text{[unit: s]} \tag{4.5}$$

It is important not to confuse between the coherence time (T_c) and the coherence bandwidth (B_c) of a channel. While the coherence time points to the variation of a channel over time, the coherence bandwidth expresses the variation of the channel over frequency and is related to the RMS delay spread of the channel.

Note: For a more extensive description of the coherence time, see [Rap02].

From Equation (4.5), it follows that the coherence time T_c is a statistical measure describing the time frame over which the channel response can be considered as invariant. Based on the ratio between the coherence time and the symbol rate, the channel can be classified as a *slow fading* channel ($T_{symbol} < T_c$), or a *fast fading* channel ($T_{symbol} > T_c$) (Figure 4.6).

Fast fading implies that the characteristics of the channel may change considerably during one symbol period. Such fast changes are impossible to keep track of in a receiver. In general, the increase of the symbol error rate due to fast fading can be modelled quite accurately by raising the noise floor of the received signal. It follows that the best way to deal with errors introduced by fast fading is to actually treat them as random noise and to pass the problem to the error coding mechanism, which is optimized for this type of errors. Sometimes, the issue of fast fading is mitigated by using a space-diversity scheme. In practice, this means that the receiver is equipped with multiple receive antennas.[6] The

[6]Note that each antenna in a multi-antenna receiver has its own analog front-end. The signals of each antenna are only recombined later on in the receive chain, after downconversion or demodulation. Therefore, a multi-antenna receiver is not the same as a receiver using a phased-array antenna, where the RF-outputs of the antennas are combined with each other in analog way to produce a directive sensitivity pattern. This does not exclude that both techniques are combined. It is perfectly possible to use a single phased array to obtain several virtual antennas looking in different directions at the same time, and then feed the signal of each virtual antenna into the different inputs of the multi-antenna receiver.

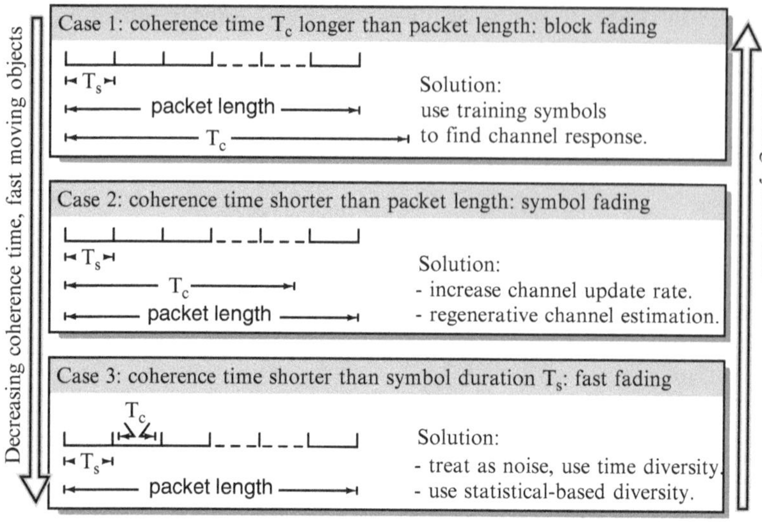

Figure 4.6. The ratio between the coherence time (T_C) of the channel and the symbol period of a transmission (T_S) determines whether the receiver operates under fast or slow fading conditions.

diversity scheme on which multi-antenna receivers are based assumes that an uncorrelated multipath scattering process takes place at each of the antennas. As such, the way in which a multi-antenna receiver deals with fast fading is rather based on statistical probabilities than on some actively controlled channel tracking mechanism. A more detailed discussion of this topic is delayed until the next section.

In practice, fast fading only occurs when the symbol rate is very low. Most data transmission applications use a symbol duration which is below the channel coherence time. They are thus operated under slow fading channel conditions, which does not necessarily imply that time varying channel parameters do not form a problem for them. The reason for this is that the estimation of the channel response is updated only once in a while, for example during the synchronization preamble at the start of each transmission burst.[7] Based on this consideration, the slow fading subdivision should be split further into symbol fading and block fading channels [McE84]. In the latter case, the coherence time is sufficiently long so that the channel response can be considered invariant during the entire period of at least one transmission burst. This is the case

[7] In the GSM system, a fixed training sequence is used. This 26-bit training sequence is located in the middle of each transmission slot, to provide the best channel estimation for both the first and the second part of the received packet.

in a static environment, where both the transmitter and the receiver are at a fixed location or are moving only very slowly.

However, in a more or less dynamic multipath environment, the most common type of slow fading is represented by symbol fading: while the impulse response of the channel may be quite stable over a period of several tens or may be hundreds of transmitted symbols, the state of the channel which was captured during the training burst of the packet will not provide an adequate accuracy over the entire time frame of a long packet burst. A possible solution for this would be to increase the refresh rate of channel response estimations by spreading several training sequences over a single packet. The refresh frequency must be chosen in order to obtain a good balance between the channel estimation overhead and the accuracy of the estimates. Another possibility would be to start with a fixed (i.e. known by the receiver) training sequence and then switch over to a regenerative channel estimation (RCE) method. In a sliding-window RCE-based approach, the decoded symbols themselves instead of the training sequence are used to make small but frequent updates to previous estimates of the channel's transfer function. The response time of the RCE-technique depends both on the frame length of the sliding window and the cut-off frequency of the noise shaping filter used by the algorithm. In order to keep track of the changing channel, either the number of symbols in the scope of the window or the cut-off frequency of the filter should be adjusted in accordance with the coherence time T_c. Either way, there is a trade-off between the speed at which updates become available and the accuracy and frequency resolution of these channel updates.

If the stability of the channel is further reduced to the duration of only a few symbols, it will become impossible for the receiver to retrieve useful information about the response of the channel: there is no way to discern between the random changes in the phase rotation of a multipath component and the modulation of the signal itself. This brings us down again to the case of a fast fading channel, where it has been already suggested that the failure to track the channel response leads to a virtual increase of the noise floor of the signal. For devices working under such conditions, the only recourse is to rely on a form of diversity that is based on statistics. This is the topic of Section 4.4, in which the use of statistical multipath diversity techniques is discussed in more detail.

Case study: impact of the coherence time on adaptive bit-loading

The main implementation issue concerning the adaptive bit-loading concept of OFDM has not been adequately addressed until now. The problem is that adaptive bit-loading requires an active intervention from transmission side: the transmitter must adapt the information load of every single subchannel according to the signal quality of that particular subchannel as seen by the receiver. Unfortunately, telepathy is one of the things the transmitter does not have at his disposal, so in a certain sense the receiver is obliged to keep the transmitter informed over the current status of the channel. The most convenient way for the receiver to do this is to release regular updates on the status of the channel. The updates can be send in the first available transmission slot or are piggybacked onto the next outgoing data packet. For a static channel of which the frequency response varies slowly over time, the feedback system is able to keep track of changing channel conditions and can avoid transmission over subchannels with an insufficient SNR.

The real problem emerges when the transmitter or the receiver are moved or are being operated in a non-stationary environment. If one or more of these conditions are met, it follows that the coherence time will shorten. As a consequence, by the time the transmitter is ready to send the next OFDM packet to the receiver, the adaptive bit-loading mechanism is already acting on outdated information.

For example, in a fairly static WLAN environment, with a receiver moving slower than 5 km/h (3 mph), measurements have shown [Sib02] that the coherence time is better than 10 ms. Suppose that the 802.11a/g system transmits 10 OFDM symbols in a single transmission burst. Taking into account the minimum interframe spacing (SIFS/DIFS) [Gas05] and the length of a single OFDM symbol including the cyclic prefix, it follows that the minimum time frame between two consecutive packets from the transmitter is 570 μs. An 802.11a/g system using the adaptive bit-loading technique should thus be able to cope with a coherence time of 10 ms.

However, these calculations were performed under the assumption that no packet collisions have been occurred and in the mean time, and no other competing systems were able to claim the next transmission slot. Also, for this example, a very low number of 10 OFDM symbols was chosen as the payload of one packet burst, which generates a lot of overhead data (e.g. the synchronization preamble). The situation becomes even worse when the receiver is moving at speeds above 35 km/h (22 mph), which reduces the coherence time even further to less than 1.37 ms [Sib02]. It should be clear that under such circumstances, the slow feedback channel between receiver and transmitter won't be able to keep up with the rapidly changing environment. The only way left for an OFDM-based system to cope with a block fading channel is to avoid the slow feedback channel in the first place and to use a combination of FEC, receiver-side channel equalization and BICM to prevent that high concentrations of errors emerge at the input of the decoder.

4.4 Multipath resolvability and link reliability

Whether the issue of multipath is approached from the side of the frequency domain or from a time-domain point of view, the outcome will always be the same: if several signal components arrive within a single sample period,[8] the receiver will not be able to discern between these individual multipath components. In more general terms: the multipath resolvability – which is the number of different paths n that a generalized, ideal receiver can distinguish – is defined by the number of symbol periods over which a multipath channel spreads the energy of a single transmitted symbol (4.6):

$$n = \text{round}\left(\frac{\tau_{\text{RMS}}}{\text{T}_\text{s}}\right) + 1 \tag{4.6}$$

It is very important to realize that the superposition of signal components that arrive within the time frame of a single symbol still causes constructive or destructive interference. In the frequency domain, this finding manifests itself as a flat fading component that is superimposed on the frequency-selective nature of the channel (which is caused by ISI). Averaged over a sufficiently long period of time,[9] no additional power is lost due to the effects of this flat fading component. However, under static channel conditions, this will cause unacceptable long outages of the link, something that cannot be solved by the time-diversity of error coding.

A generalized RAKE receiver architecture increases the reliability under static channel conditions by combining the energy of the resolved multipath components.[10] Seen over a longer period of time, the individual paths that can be resolved by the receiver experience – independently from each other – Rayleigh or Rician frequency flat fading.[11] While the reliability of a single resolved path is not better than what is achieved in a non frequency-selective flat fading channel, combining the energy of several of such independently fading multipaths increases the reliability of a communication link. It follows that intersymbol interference is a necessary evil for a reliable wireless link in the multipath environment.

---

[8] If $\tau_{\text{RMS}} < 10 \cdot \text{T}_\text{s}$, then we have a flat fading channel.

[9] This depends on the coherence time T_c of the channel.

[10] There are several ways to combine the energy of multiple symbol streams. The most common technique is *selection diversity combining* (SDC). A good example of SDC is switching between two antennas and selecting only the antenna which offers the best signal quality. More advanced techniques combine the energy of multiple resolved streams before the demodulator takes a decision about the value of the received symbol. These techniques are known as *maximum ratio combining* (MRC) or *equal gain combining* (EGC). The latter one is in fact a simplified version of MRC, but ignores the signal-to-noise ratio of the different multipath components. More information on this topic can be found in [Fuj01].

[11] Rician fading is the model for line-of-sight (LOS) fading, while Rayleigh fading if the stochastic model for non-LOS fading [Pro00].

Of course, the channel must offer a sufficient amount of resolvable multipath components. In the heavy multipath indoor channel, increasing the multipath resolvability simply comes down to the point that the symbol duration must be chosen sufficiently shorter than the RMS delay spread of the channel. If the correlation between the resolved paths is sufficiently low, the individual symbol streams can be assumed to have independent fading characteristics [Cha79]. While the total signal power integrated over all resolved components still has a statistical time-varying fading profile, the reliability under static channel conditions is improved. This statement can be proven as follows.

Figure 4.7 shows the simulated PDF of the received signal power in case of a single resolved Rayleigh fading NLOS path. First, note that the signal envelope of a Rayleigh distributed NLOS channel has a large peak at lower values and a long tail for higher signal magnitudes. Relative to the average power that is available from the channel, this means that the link suffers from destructive interference during considerable periods of time. However, how does this finding fit with the previous conclusion that no power (and as a consequence also no Shannon capacity) is lost when the channel is evaluated over a longer period of time? The answer here is given by the long tail of the Rayleigh distribution: from time to time, it happens that a considerable number of multipath components interferes in a constructive manner. At such occasions, a very high signal-to-noise ratio will temporarily boost the information capacity of the channel. At least theoretically though. In any real-world receiver implementation, the

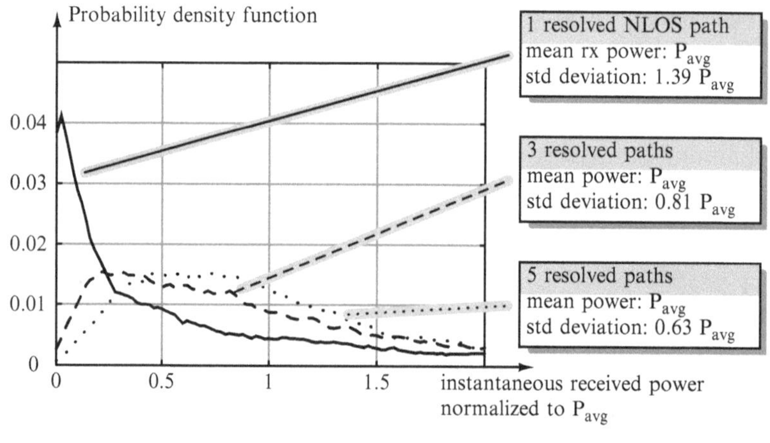

Figure 4.7. Simulated probability density for the instantaneous received power with respect to the average received power (P_{avg}) in a Rayleigh-fading NLOS channel. A higher number of resolved multipath components results in a reduced standard deviation.

modulation depth that would be necessary to achieve this goal can simply not be met. In practice, the inability of hardware to keep up with the wide dynamic range of possible channel conditions is eventually translated into either a higher transmission power or a reduced information throughput.

Continuing the example of the NLOS channel, suppose that the delay spread of the channel allows the receiver to resolve n multipath components, instead of the single path from above. Remember that each of the resolved symbol streams still suffers from (independently) Rayleigh fading as a result of interference within a data symbol. The benefit of a RAKE receiver becomes clear when the signal power of several of these multipath components is combined to form a new symbol stream. The combination of n independent Rayleigh-distributed random variables – in this particular case the vector magnitude of the different resolved symbol streams – results in a new randomly distributed variable. The probability density of this variable can be calculated using the n-fold convolution of the PDF's of the summed variables [Lun74]. The mathematical expression for the sum of n random distributed variables is not straightforward to compute and is out of the scope of this text. However, a few interesting characteristics that are worth to be underlined.

First of all, the sum of two Rayleigh-distributed random variables does not have a Rayleigh PDF any more. According to the central limit theorem [Fel45], the sum of independent arbitrary distributed random variables will approach a normal Gaussian distribution if a large number of such variables is combined. The combination of only a few Rayleigh distributed variables holds the middle ground between a Rayleigh shaped and a Gaussian shaped distribution. Also, it turns out that the point of maximum population density shifts from a rather low value as a result of the Rayleigh-like distribution, to the mean value ($\mu = 1$) when more resolved multipath signal components are combined and the PDF shape becomes more Gaussian (Figure 4.7).

In order to make a comparison between RAKE and non-RAKE receivers, the total combined energy from all multipath components has been normalized to the average received power. The rationale behind this is that the received energy per symbol must remain constant (E_b/N_0 = constant) for a fair comparison. In a NLOS environment were all paths approximately share the same average power, resolving n multipath components implies not only that the average total power is divided over those n resolved paths, but the standard deviation of the power in a path becomes also n times smaller. Many engineers know by heart that the variance of the sum of uncorrelated random variables is given by the sum of their variances [Bei01], so the following expression for the standard deviation of the combined signal power (not magnitude) at the output of a RAKE combiner holds (4.6):

$$\sigma_n = \sqrt{\left(\frac{\sigma_1}{n}\right)^2 + \left(\frac{\sigma_2}{n}\right)^2 + \cdots + \left(\frac{\sigma_n}{n}\right)^2}$$
$$\approx \frac{1}{\sqrt{n}}\,\sigma_{\text{Rayleigh}}\,, \tag{4.7}$$

where n is the number of independently Rayleigh fading multipath streams that are being resolved in the receiver and σ_i is the standard deviation of the power of the i-th multipath component.

It follows that the deviation from the mean power that can be extracted from the channel becomes smaller when the number of resolved paths is increased. In other words, the number of occasions where the signal power drops below a certain minimal level is drastically reduced. This is great news for the reliability of the wireless link in a static environment.

For example, suppose that the design specifications of a certain wireless link prescribe a 90% uptime guarantee, while the remaining 10% outage time is absorbed by the time-domain diversity of the error coder. Figure 4.8 shows the cumulative distribution function (CDF) of the power distribution for different numbers of resolved paths in a NLOS channel. A receiver unaware of the 'multipath' concept shows a crossover value of 0.01. This actually boils down to

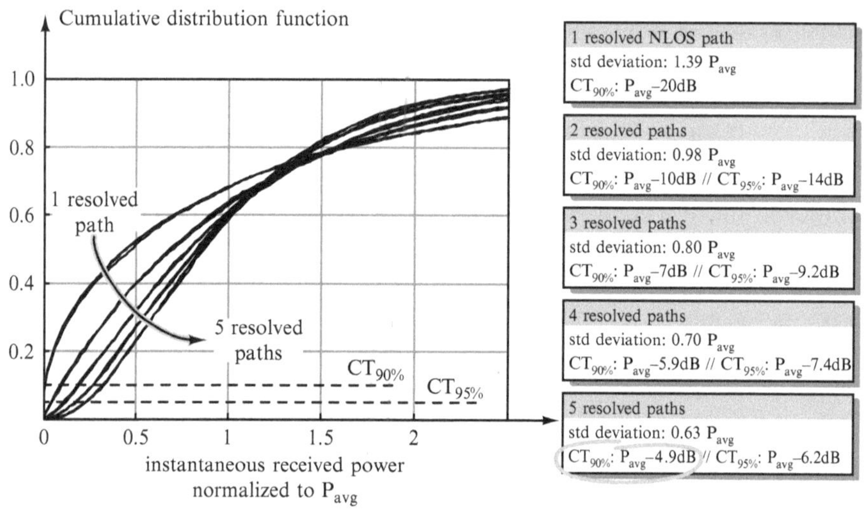

Figure 4.8. Confidence thresholds ($\text{CT}_{90\%}$ and $\text{CT}_{95\%}$) for a NLOS channel and one to five resolved multipath components. For example, in case of five resolved paths, the instantaneous power available to the receiver is higher than $P_{avg} - 4.9$ dB for 90% of the time.

the fact that the average signal power that is available at the output of the channel should be at least 20 dB (100 times!) higher than the sensitivity level of the receiver. For five resolved multipath components, only 4.9 dB of the link budget must be spent to achieve the same 90% reliability requirement.

As a final remark, it is stressed that the fading problem discussed here is independent of the problems related to frequency-selective fading originating from ISI: in the discussion above it was supposed that the transceiver system did perfectly separate all resolvable multipath components.

Case study: reliability issues of 802.11a/g

One of the main advantages of using OFDM as the modulation method in a wideband communication system is that the problem of ISI is easily circumvented by splitting the frequency band in multiple subchannels. Each of the subchannels is then modulated at a symbol rate which is much lower than the RMS delay spread of the channel. In the 802.11a/g system, for example, the duration of a single OFDM symbol block (4 μs) is guaranteed to much longer than the delay spread in a typical indoor environment (10–70 ns). The rate at which the data samples[12] are modulated is 50ns. At the antenna terminal of the receiver, the energy related to one transmitted symbol is thus spread over two or three adjacent data symbols.

In theory, the 802.11a/g system is thus able to combine the signal power of two or three multipath components. This should not be interpreted as if there were a bundle of individual symbols streams at the output of the OFDM receiver: it is the sum of the signal power from all resolved multipaths that is equal to the total power available from all subbands. Apart from the issue of ISI, this implies that if the number of virtually resolved multipath components becomes higher, the deviation on the total power in all OFDM subbands becomes lower and the system becomes more reliable. Since the RMS delay spread of 802.11a/g is limited to only three or less symbol periods for the typical indoor channel, the reliability of the system is doomed to collapse in NLOS fading channels that lack a dominating multipath signal component: it follows from Figure 4.8 that 7 up to 10 dB of the total link budget is irreversibly lost under static NLOS conditions. In practice, this means that the transmitter is forced to reduced the modulation depth in order to obtain a more reliable value for E_b/N_0. From Table 1.1 (p. 16), it can be verified that a loss of 10 dB is equivalent to a reduction of 4 in raw signal throughput, or a link failure when the modulation depth cannot be reduced any more. Note that this problem will arise both in the adaptive as well as in the bit interleaved (BICM) bit-loading schemes because it

[12]An OFDM symbol is composed of 64 complex data samples preceded by a 16-sample cyclic prefix. The 80 data samples with a duration of 50 ns result in the OFDM symbol period of 4 μs [Wla07].

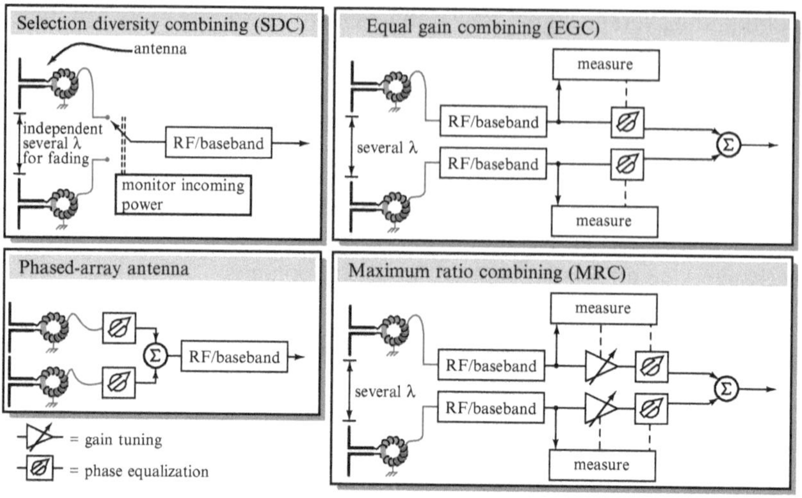

Figure 4.9. Spatial diversity improves the link reliability over fading channels. Diversity combining should not be confused with phased-array antennas with a directive sensitivity pattern.

originates from the lack of RMS delay spread rather than from a malfunctioning of the subchannel loading mechanism.

It is for this reason that spatial diversity has recently regained considerable interest as a reception enhancement technique for OFDM-based WLAN receivers. The importance of this form of diversity was already recognized in the early 1970s [Dav71], but only recently found widespread use thanks to the availability of digital signal processing. In order to combat the dreadful misery caused by destructive intra-symbol interference, a spatial-rake receiver combines the energy from different signals collected by two or even more antennas.

As with any other problem, your mileage may vary depending on the complexity of the solution. The most widespread implementation of an multi-antenna receiver employs a *selection diversity combining* (SDC) scheme. In this scenario, the signal quality at each antenna is monitored while only the antenna which offers the highest SNR is retained for reception (Figure 4.9). Because of its simplicity, selection diversity is used in many 802.11a/b/g devices. The performance of SDC is somewhat limited, of course, since this kind of receiver will not be able to use all the signal power that is available from both antenna terminals. The only reason for the success of SDC is that it does not require a duplication of (part of) the receiver front-end hardware, apart from the physically separated antennas.

Chapter 5

PULSE-BASED WIDEBAND RADIO

Recently, a lot of attention went to a decision [Com02] of the Federal Communications Commission (FCC) to unblock 7,500 MHz of spectrum in the 3.1 to 10.6 GHz frequency band (Figure 5.1). Of course, this has generated a lot of interest from both industry and academics, since the enormous amount of bandwidth provides a lot of perspectives and new opportunities for broadband data communication applications, or the so-called *Ultra-Wideband* (UWB) systems. From a marketing point of view, devices that are able to use such a large amount of spectrum will become the perfect replacement for the video cable, wireless LAN and are at the same time the 'enablers' for new technologies with odd names such as there is the *cable-free universal serial bus* (cable-free USB) and the even more remarkable *wireless firewire*.[1] However, things are not always as they seem to be and have to be put in perspective.

First of all, the maximum EIRP[2] which an UWB-transmitter is allowed to use is very low. The limit on the average power spectral density is −41.3 dBm/MHz, which comes down to −2.55 dBm (about half a milliwatt) for the entire 7.5 GHz block. Wideband radio may offer interesting possibilities, but the expectations must remain down-to-earth under the given circumstances. Using the numbers given above, it is possible to verify this with some rough calculations. For example, suppose a wireless system which uses a spectral bandwidth of 500 MHz. Also, the system is being operated in an ideal channel without multipath reflections.[3] Under the assumption that there is no interference and

[1] Not to confuse with toothless Bluetooth.

[2] EIRP: Effective Isotropic Radiated Power.

[3] Path loss calculated using the Friis antenna equation in free space, center frequency 5 GHz [Poz05].

W. Vereecken and M. Steyaert, *Ultra-Wideband Pulse-based Radio: Reliable Communication over a Wideband Channel*, Analog Circuits and Signal Processing Series,

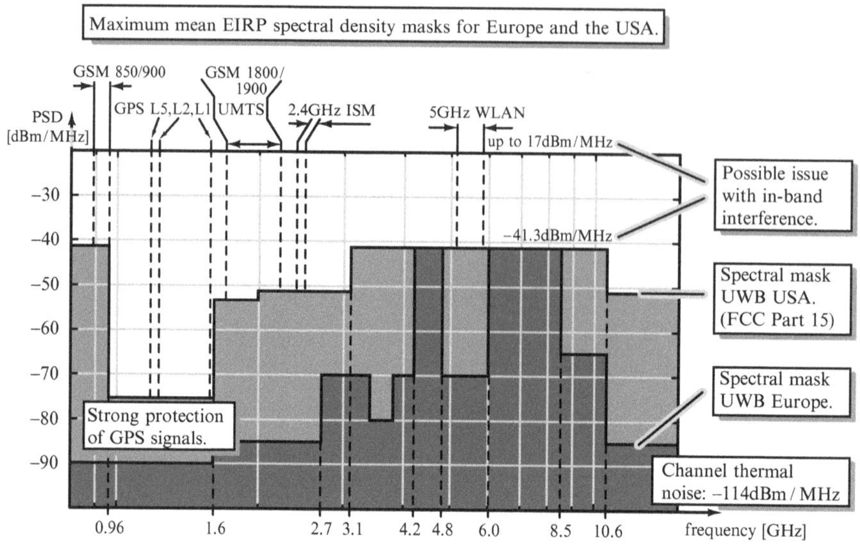

Figure 5.1. Spectral power mask for UWB communication devices below 10.6 MHz. Note the strong protection of GPS signals and the imminent interference problem caused by high-PSD wireless LAN radio transmitters in the 5.150–5.825 GHz band.

only thermal noise is present in the channel, its theoretical Shannon capacity (C) at a link distance of 10 m is given by (5.1):

THEORETICAL THROUGHPUT OF AN UWB-LINK (10m)

$$\begin{aligned} \text{Pathloss}_{@10\text{m}} &= 47\,\text{dB} + 3\cdot 10\log_{10}(10\text{m}/1\text{m}) = 77\,\text{dB} \\ \text{Signal power} &= -41.3\,\text{dBm/MHz} + 10\log_{10}(\text{B}) - \text{PL}_{@10\text{m}} \\ \text{Noise}_{290\text{K}} &= \text{kTB} = -114\,\text{dBm/MHz} + 10\log_{10}(\text{B}) \\ \text{SNR} &= -41.3\,\text{dBm/MHz} + 114\,\text{dBm/MHz} - \text{PL}_{10\text{m}} = -4.3\,\text{dB} \\ \text{B} &= 500\,\text{MHz} \\ \text{C} &\approx \text{B}\log_2(1+\text{SNR}) = 228\,\text{Mbit/s} \end{aligned} \tag{5.1}$$

In a standard receiver architecture, the large spectral footprint of a wideband radio system has some important consequences that are easily underestimated. Most digital communication systems, such as 802.11a/b/g, employ a higher order modulation scheme to achieve a fairly high information density and throughput in a compact frequency band. For the low-PSD wideband radio architectures discussed in this section, the situation is exactly the other way around. First of all, remark the low average SNR of −4.3 dB in (5.1). The spectral efficiency of the system is below 1 bit/s/Hz and it is only thanks to the large amount of available bandwidth that a decent data throughput can be achieved. However, the low modulation depth used by a low-PSD radio system does not necessarily lead to an equivalent reduction of the dynamic range in the front-end of the receiver. The problem here is caused by the fact that the linearity requirements are not dictated by the dynamic range of the signal-of-interest itself, but by the power level of in-band interferers. It is the combination of a high bandwidth and dynamic range that forms the biggest challenge for wideband radio receivers. On the other hand, even when every technical requirement has been met, the capabilities and power consumption of the receiver may not be reflected in a higher throughput of the system: after all, the signal-of-interest is still below the noise floor of the channel.

In the case of OFDM-modulated wideband radio, for example, a lot of effort is put in obtaining an accurate digital representation of the entire frequency band. Because the actual signal-of-interest is only a small portion of the total information that can be represented by the digitized OFDM symbol, this will inevitably result in a large amount of overhead data. This is because in OFDM, both modulation and demodulation are performed in the digital domain. By design, the transmitter of such a system is forced to represent the entire frequency band in a mathematical form before it can be converted to the analog domain. The same reasoning holds for the receiver. This time, there are some additional difficulties which are caused by interference and the linearity of the input stage. The only remaining task for the analog processing circuits (which includes the DA/AD converter) is to convert between a baseband digital representation and the passband-RF analog form of the time-domain OFDM signal.

The particular problems that are faced in a wideband low-PSD radio system (linearity, sparse spectral information density and in-band interference) suggests that trying to maintain control over the complete (de)modulation process in the digital domain at all costs may not result in the most power efficient system. It is the purpose of this section to introduce an architecture for wideband radio systems in which part of the modulation process is outsourced to the analog domain. The presented approach tackles the problems caused by the nature of wideband radio at their roots. Some part of the signal processing will still reside in the digital domain, because of its flexibility and the ability of digital processing to exercise active control over the analog subsystem. However, for efficiency reasons, a considerable amount of signal processing will thus occur in the analog domain. As will become clear in the following sections, the analog subsystem is not only used as a dumb interface between the digital processor and the analog outside world (i.e. the antenna terminal), but it also plays an important role in the *bandwidth compression*[4] of the RF antenna-signal.

[4]The clearest example of bandwidth compression is the conversion from passband to baseband.

The presented wideband radio architecture is based on a single-carrier modulation method, not only because of the above mentioned reasons, but to avoid in advance all difficulties that could emerge from the high Crest factor[5] (CF) of multicarrier modulation. Thanks to the use of the interferer suppression and signal reconstruction method (ISSR) introduced in Section 5.3, the system will be able to survive in hostile environments with in-band interference and multipath reflections. The performance of the presented system will be a match for an OFDM-based system, but at only a fraction of the power consumption.

5.1 Symbol rate versus multipath resolvability

It was shown before (Section 4.4) that a system using a high symbol rate has a high multipath resolvability if a RAKE-compatible[6] system architecture is used. The ability to resolve multipath signals is an important aspect of a wireless system since it can help to increase the reliability of the system in slow fading channels. However, resolving multiple reflections of the same signal becomes increasingly difficult in the indoor environment, where the received signal has a very limited delay spread, from less than 10 ns up to a very few 100 ns. Increasing the symbol rate only to increase the resolvability of the system may not be a good investment, for several reasons. First of all, increasing the symbol rate will increase the bandwidth of the entire transceiver and receiver chain, with the obvious consequences for the power consumption. In low-PSD radio applications (e.g. UWB), the extra effort does not automatically translate into an acceptable performance in terms of throughput. The reason for this is that the symbol rate can become higher than the theoretical capacity of the channel. Of course, this issue is easily resolved by combining multiple data bits (also named *chips*) into a single bit of information, but there are some important side-effects which are easily underestimated. Increasing the input bandwidth of the receiver not only makes the signal-processing stages more vulnerable to accidental high-power narrowband interferers, but also leaves the door wide open for the background (thermal) noise in the channel. As a conclusion, increasing the symbol rate in order to increase the reliability of the system may cause more problems than it solves. It is this insight that leads us to the idea of pulse-based wideband radio.

In a pulse-based radio system, the symbols which are transmitted through the channel are represented by short pulses (Figure 5.2). The duration and shape of the individual impulses determines the spectral footprint of the pulse-based transmission. This in sharp contrast to the continuous-time modulation

[5]Crest factor: $\mathrm{CF} = \sqrt{\mathrm{PAPR}}$.

[6]RAKE-compatible: any receiver which is able to disentangle the different multipath streams. This can be in the time domain (e.g. DSSS receiver) or by an OFDM-alike system which is immune to ISI.

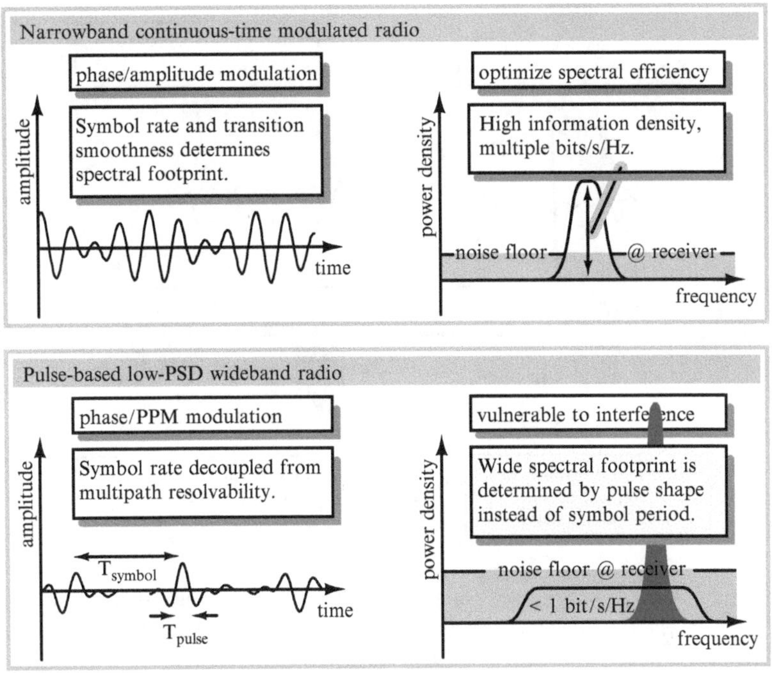

Figure 5.2. A pulse-based radio system decouples symbol rate from multipath resolvability. However, the wideband character of the transmission makes it very vulnerable to in-band interference.

approach in single- and multicarrier systems, where the minimum transmission bandwidth is determined by the symbol rate itself. Moreover, in the latter case, no stone[7] is left unturned in the pursuit of maximum bandwidth efficiency.[8] By making the pulse period (T_p) smaller than the interval between two symbols (T_s), a pulse-based radio decouples the symbol rate from the duration of the transmitted pulses.

This has significant consequences for the multipath resolvability of the system: even if the symbol rate is lower than the coherence bandwidth (B_c) of the channel, it is theoretically possible for a receiver to acquire multiple samples of the channel and resolve different delayed versions from the same pulse with a time accuracy which is higher than the symbol rate T_S. Separately resolved pulses can afterwards be joined together in order to reduce the variance on the combined signal-to-noise ratio as seen by the receiver. In a standard continuous-time modulated system, merely increasing the sample rate of the

[7] Stone: conceptual metaphor for sidelobe. Also known as a small rock outside this context.

[8] In a continuous time transmission system, the spectral mask of the symbols is commonly restricted without introducing ISI by using a raised cosine rolloff filter.

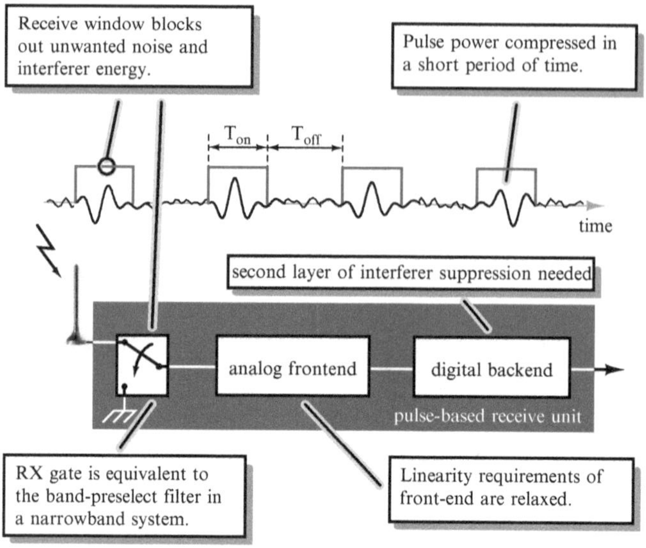

Figure 5.3. In the pulse-based radio concept, the transmitted energy is compressed in short pulses. This allows the receiver to block interferer power when no pulse is expected to arrive.

receiver (while maintaining the transmitted symbol rate) does not improve the reliability of the wireless link. This is caused by the fact that if the receiver takes multiple samples within the period of a single transmitted symbol, the assumption of independently fading samples does not hold any longer. Combining the energy of several resolved multipath components will thus not reduce the effects of fading in order to produce a more reliable signal. Concluding, Equation (4.6) does not apply for dependent fading signals.

Apart from a better multipath resolvability in indoor channels with a short delay spread, the concept of using short pulses instead of a continuous-time modulated system has some additional advantages. In a pulse-based radio system, the transmitted energy is compressed in a short period of time (Figure 5.3). This implies that, if the receiver is aware of the period between two consecutive transmitted pulses, it can anticipate on this by only allowing RF-energy from the antenna to enter the front-end when a pulse is expected to arrive. This technique can be easily extended for a RAKE receiver architecture: only when a certain multipath component is expected to arrive at the antenna, the front-end is activated. For the remaining time, all RF-signals arriving at the antenna terminal are blissfully ignored. The result is that a receiver which employs this *gating* or *windowing* technique, will be less susceptible to the increased channel background noise caused by the large input bandwidth of the receiver.[9] The same holds for accidental in-band narrowband interferers: they are suppressed by the ratio between the on- and off-state (duty-cycle) of the receive window. This implies that if the gating switch is located nearby the antenna terminal, the share of interferer power in the total received power figure is effectively reduced. One might think about this receive window as the time-domain equivalent of a channel-select filter in the frequency-domain. The result is the same, though, as unwanted interferer power is blocked early in the receive chain, and the signal-of-interest is passed unattenuated to the subsequent input stages.

For example, if the total on-time of the window adds up to 10% of the average interval between two pulses, an unwanted narrowband interferer will be suppressed by 10 dB. For an OFDM-based radio approach, the unattenuated interferer will have to pass through the entire receive and AD conversion chain before it can be suppressed in the digital back-end of the receiver. As noted earlier, the receiver puts a lot of effort (i.e. power) in the pursuit of an accurate representation of this interferer in the digital domain, only to get rid of it at the first available opportunity. Of course, an interferer suppression of only around 10 dB is far from sufficient for any practical application. However, the pulse-based receiver architecture which will be introduced later on in this section employs a two-level interference suppression scheme. Most of the interferer energy is suppressed by the coarse window switch early in the receive chain. This greatly relaxes the linearity requirements of the successive analog front-end. In a second step, the remaining unwanted spurious energy is removed by a more flexible, fast-adaptive filter in the digital domain.

In the discussion above, the outlines of a pulse-based radio system have been described, but only very superficially. As the attentive reader may have guessed, the benefits of using pulses instead of continuous-time modulated symbols do not come for free. Two major problems have to be solved before such an unconventional radio system can be deployed.

Window alignment

First, there is the problem of getting the receive window in line with the arrival time of the pulses at the antenna. Failing to do this results in a complete signal loss, since a misaligned window must be regarded as the time-domain equivalent of a channel-select filter that is tuned into the wrong frequency band. The shorter the duration of the pulses, the higher the gain in terms of signal-to-noise ratio, but the more difficult it becomes to 'tune' into the correct window position. As in many situations, there is always a trade-off between

[9] In a narrowband *subsampling receiver*, exactly the opposite effect is experienced. The reason for this is that a subsampling receiver rejects some valuable signal power, as a result of which the SNR is decreased.

performance, cost and power consumption: the duration of the locking process depends on the number of possible window positions in the time frame between two consecutive pulses, and the number of positions that the receiver can observe in parallel. In practice, the number of parallel receive paths is kept very limited, due to the considerable cost in terms of silicon area. So the only parameter left to play with is the duration of the window: increasing the duration of the on-state of the receive window reduces the number of window locations and reduces the synchronization time overhead, but it also causes a proportional increase of the noise and interferer power that is captured by the receiver. A compromise would be to start the synchronization process with an increased window length. After the receiver has acquired lock on one of the multipath streams of the transmitter, the signal-to-noise ratio can be improved by a stepwise decrease of the window length until the optimal result is achieved (Figure 5.4). Of course, this implies that the receiver must have disposal over at least two parallel receiver units, since the continuous operation of the link has to be guaranteed during the quest for better signal quality. A more detailed analysis of this topic is covered later in this section.

Finding the location of the pulses is one thing, but after the receiver has acquired a successful lock, it should be able to track the accumulating time error

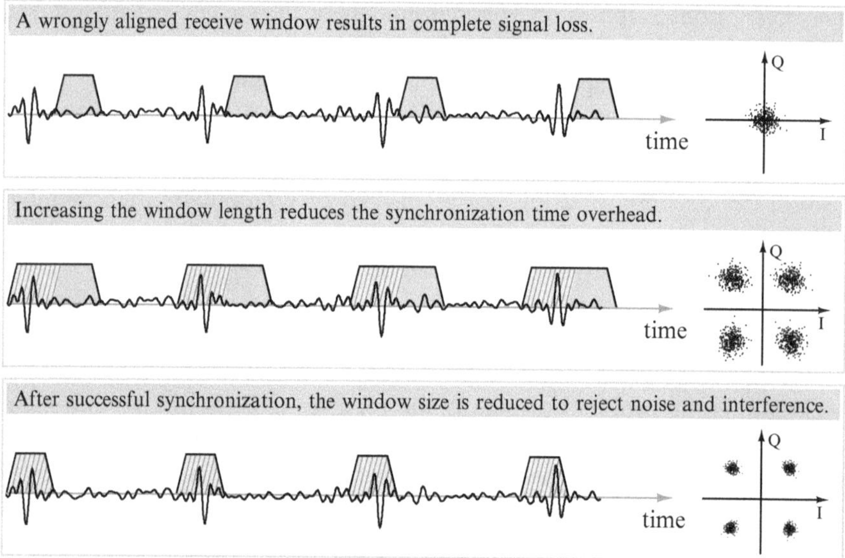

Figure 5.4. The receive window in a pulse-based radio is the equivalent to tuning the band-select filter in a narrowband architecture. Note that this figure is only a simplified representation of reality, since the effect of multipath reflections is omitted in this example.

between the received pulses and the window position controlled by the receiver: a small frequency offset between the clocks of the transmitter and the receiver will cause an increasing error between their time references. If the receiver is unable to monitor these offsets, this eventually will result in signal loss from the moment the pulses shift outside the scope of the receive window.[10]

For example, if the clock offset between transmitter and receiver is 25 ppm and the duration of the receive window is set to 1 ns, it will take around 40 μs for the pulses to go completely out of scope. For a symbol rate of 100 Msymb/s, the tracking algorithm has a time frame of 4,000 pulses to track and compensate for this error. Despite the fact that in a pulse-based radio system the symbol rate is decoupled from the pulse duration, this finding shows that there is a physical constraint on the ratio between the pulse duration and the symbol interval: for a very low bit rate, the number of symbols before the pulses become out-of-sync with the receive window is too limited for the receiver to be able to obtain a reliable estimate on the clock offset. A possible solution for these low data rate communication systems could be to use a sufficiently high symbol rate and to transmit data in short bursts. But the reader should be aware that if the idea behind using a low data rate is to save power, the use of a pulse-based wideband radio system may not be a good design choice.[11] After all, a pulse-based receiver must keep the entire wideband front-end running in order to monitor the channel for incoming messages.

Bandwidth compression: the basis of power efficient pulse-based radio

The second issue that has to be addressed by the designer of a pulse-based radio receiver is the considerable bandwidth of the received pulses. The advantage of using short pulses (typically 2 ns or less) instead of a continuous-modulated signal is that the symbol rate is independent from the spectral characteristics of the transmitted signal. The use of short pulses provides an improved multipath resolvability in indoor channels with a low delay-spread, without the need to increase the symbol rate to impractical high levels. In terms of power consumption, the big bonus of using pulses is that a significant amount of interferer power can be eliminated early in the receive chain.

The benefit in terms of bandwidth depends on where the demodulation of the pulses is done. If demodulation and symbol demapping is deferred to the digital

[10] The frequency offset of an XTAL-based oscillator is typically a few tens of ppm.

[11] One of the applications where the use of a low-throughput wideband system could be justified is the use in RFID asset tracking tags [Fon04]. The tags themselves do not have their own power supply (under a best-case scenario a small battery) and the tag contains only a small pulse-based transmitter. The use of short pulses allows a network of receivers to collect ranging data – much like a radar system – in order to determine the approximate location of a tag.

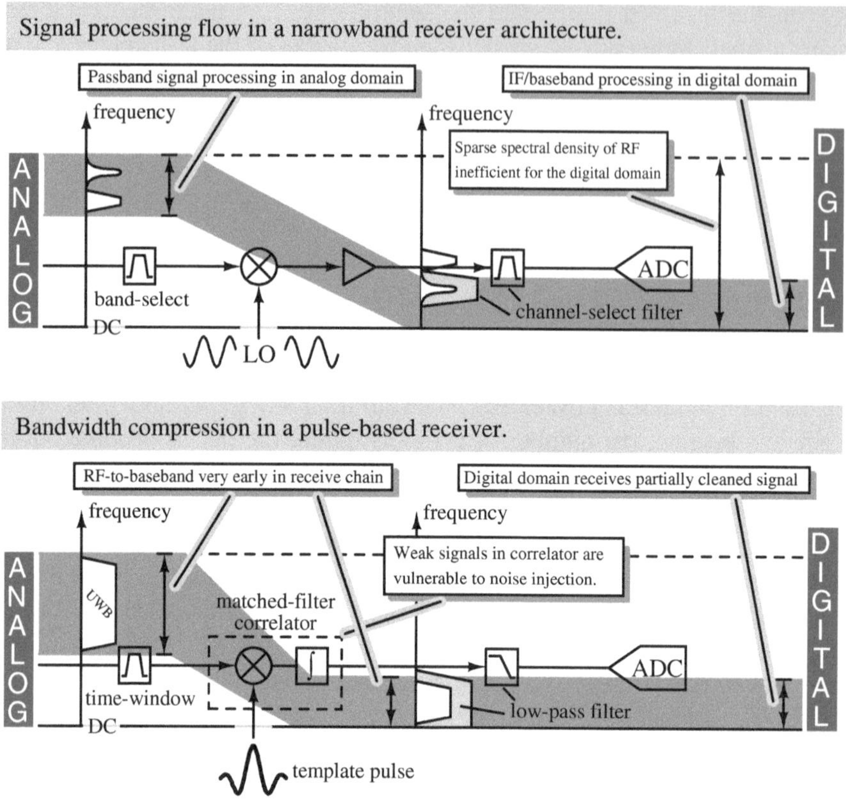

Figure 5.5. Analogously to the approach used in a narrowband system, the wide spectral footprint of pulse-based radio must be compressed to a limited-bandwidth baseband signal as soon as possible in the receive chain.

back-end, then the entire receive chain must support the large bandwidth of the RF-signal. A significant amount of power can be saved by moving (part of) the demodulation process to the front of the receive chain. In a pulse-based radio system, the demodulation or detection of the short pulses can be seen as the equivalent of downconversion in a carrier-based transmission (Figure 5.5). In the downconversion process, the RF-signal is translated from a passband signal – adjusted for transmission over the wireless medium – to a baseband signal. In virtually all cases, downconversion is performed by the analog domain. Unlike analog circuits, it is much more energy-efficient for a digital system to process a baseband signal than to boldly deal with the RF-signal itself. For the same reasons, a pulse-based radio receiver could employ some sort of bandwidth compression technique to convert the RF-pulses into a baseband signal.

Theoretically, the best way to accomplish this would be to use a matched filter receiver, where the input signal is correlated with a template pulse. Several strong arguments can be formulated against this approach. One of the prerequisites of using a matched filter is that the shape of the input waveform is known in advance. However, assuming that the path from the input terminal of the transmitting antenna all the way down to the receiver can be approximated by a flat frequency- and a linear phase-characteristic is a huge mistake. The waveform of the pulses depends on the matching between the transmitter (receiver) and its antenna, the geometrical radiation point of a certain frequency on the antenna surface and whether pulses overlap or not (as a result of multipath interference). Under the assumption of a LOS channel, the initial radiation characteristics of the antennas will also drastically change if an electromagnetic responsive material is brought in the neighborhood of the transmitter or the receiver.

Even if an acceptably accurate template pulse waveform could be constructed, there is still a second and potentially far more serious problem related to the demodulation of pulses using a matched filter. This time, the problem is only brought to the surface by diving deep into the circuit level of a receiver design. Remember that our original purpose was to bring (part of) the demodulation process closer to the antenna of the receiver. For example, in a heterodyne receiver, the downconversion mixer is located just behind the LNA. At this point, the signal-of-interest is still extremely weak, in the order of a few (tens of) microvolts. In order to obtain a high conversion gain, the amplitude swing of the LO-inputs[12] of the mixer is kept as high as possible. Any practical implementation will suffer from spurious frequencies at the mixer output due to LO-feedthrough and self-mixing (Figure 5.6).

Fortunately, in the case of a single frequency in the LO-signal this results in a DC-offset[13] and some spurious signals at the LO-frequency and at some higher harmonics of f_{LO} (Figure 5.6). However, doing the same in the correlation filter of a pulse-based radio system is doomed to fail catastrophically, for the following reason. The power spectral density of the template pulse waveform applied to the mixer inputs of an analog correlator spans the entire pulse spectrum. As a result of self-mixing and leakage, the output spectrum of the mixer will thus contain unwanted components over the entire range from DC to twice the maximum pulse frequency. Since the power of the locally generated template signal is much stronger than the incoming RF-signal, these spurious components could cause clipping further down the chain. Even if the pulses are spaced

[12] LO: local oscillator.

[13] This DC-offset signal is the biggest architectural problem in zero-IF receivers.

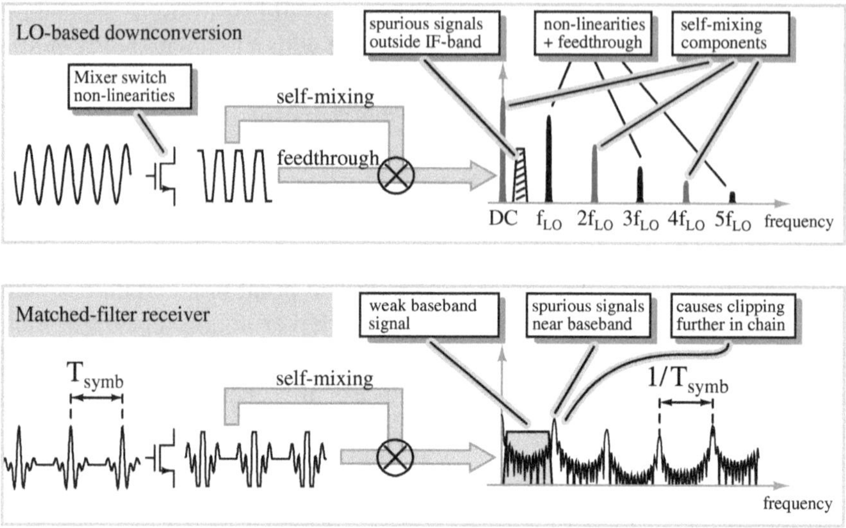

Figure 5.6. The template signal causes spurious components near and inside the signal band in a matched filter based pulse detection system. This effect desensitizes the receiver.

uniformly in time,[14] removing the unwanted spurs from the signal spectrum requires an extremely sharp filter since the nearest component is located at the Nyquist frequency of the symbol sampling frequency. As a consequence, the IF-amplifier of a matched filter receiver must be operated at some back-off point from the optimum gain value so that the receiver is desensitized. In fact, the mechanism is the same as for an externally injected narrowband interferer.

Replacing the template signal by a single LO-frequency

A general rule-of-thumb is that switching or repetitive signal components should be avoided inside the analog signal path of a receiver, or at least for frequencies that could corrupt the spectrum of the signal-of-interest. Although a matched filter may be theoretically the most optimal solution, the adverse consequences at transistor-level make it perform much worse than would be expected at system-level. The most convenient solution to the problem of self-mixing is to replace the template waveform by a continuous running single-frequency LO-signal. Doing so also avoids that the template signal must be aligned to the unknown and varying arrival time of the pulses.

[14] This results in spurious components that are located at discrete multiples of the symbol repetition frequency.

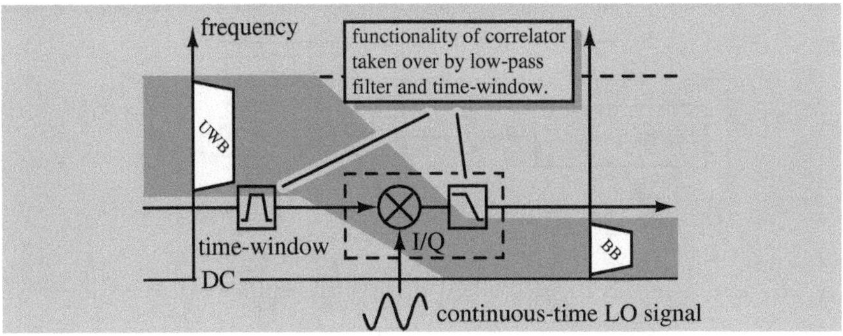

Ruling out the matched filter receiver, the reader may question if there is still any difference with a standard heterodyne receiver architecture. Indeed, the only remaining part making the described architecture suited for pulse-based radio reception is the window switch in front of the receive chain. In combination with the mixer and the low-pass filter, the switch forms a basic correlator structure and is the essential component to resolve the closely spaced multipath components. Based on the previous reflections, the reader should start to see the idea of pulse-based radio in a new light. Instead of focussing on the generation and demodulation of pulses, the pulse-based radio system that will be elaborated in the next sections should be considered as an *extension on the narrowband radio architecture*.

This cryptic statement should be understood as follows. Suppose a single-carrier QPSK modulated radio system. Without any change to its working principles, the transceiver hardware is expanded with an extra layer between the transmitter (receiver) and the channel, with the purpose to increase the multipath resolvability of the system: instead of dropping the RF-carrier of the transmitter directly into the transmission antenna, only a small fraction of the RF-signal is allowed to pass (Figure 5.7). The signal that is applied to the channel has the appearance of short pulses, but is in fact a well-controlled QPSK-modulated signal. Remark that cutting away some of the time-domain information does not alter the phase-information contained in the residual signal that is applied to the antenna. The duration of the time-window determines the spectral mask of the transmission, but apart from this, the window is not important for the QPSK symbol as such. Also important is that jitter on the edges of the window does not corrupt the phase angle of the transmitted constellation point. The window switch in the analog front-end has to be synchronized with the I/Q-modulator in the digital part of the transmitter. Transitions between two constellation points in the QPSK constellation diagram must occur during the off-state of the transmit window. The way the transition is done is of little importance, as long as the carrier phase has been settled when the transmit

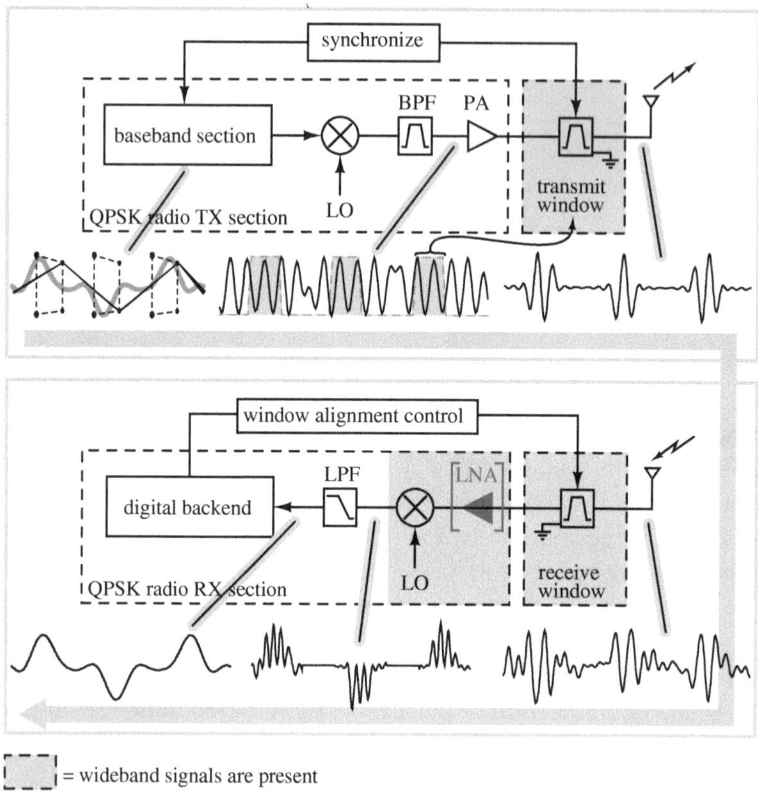

Figure 5.7. Instead of focussing on the generation/demodulation of pulses, a pulse-based radio must be considered as a regular QPSK receiver with an extension layer decoupling symbol rate from multipath resolvability.

window opens again. The same procedure is repeated exactly the same way at receiver side, where the antenna signal is only gated to the receive chain at the moment a pulse is expected to arrive.

In contrast to the transmitter, the receiver has no clue about the moment when the receive window must be activated. Finding the exact timing is part of the synchronization procedure and will be discussed in the next section. Leaving the synchronization aspect aside, consider the remaining part of the receive chain. As explained earlier, maintaining a wide bandwidth needed to represent the short pulses along the entire receive chain is a real kiss of death for low-power applications. One part of the demodulation process, which is the bandwidth compression of the pulses, should be performed as soon as possible after the RF-signal has passed the receive window switch. This time, fortunately, the explanation takes longer than the actual implementation. The bandwidth of the RF-pulses can be compressed to a baseband signal by an ordinary low-pass filter, located at the output of the downconversion mixer. From this moment on, it will appear as if a regular QPSK modulated signal is being received: because the bandwidth of low-pass filter and the consecutive analog signal processing chain is too small to represent the short pulses, the gap between two pulse symbols is automatically filled just as if a regularly continuous modulated QPSK signal is being received.

The bandwidth of the filter cannot be chosen arbitrarily, though: the information gets corrupted when the minimum bandwidth constraint necessary to represent the original baseband signal is not met. It is also clear that an analog-domain implementation of the low-pass filter will cause considerable inter-symbol interference (ISI) when the 3 dB-bandwidth of the filter approaches the Nyquist bandwidth of the baseband QPSK stream. This is due to the infinite impulse response of analog (continuous-time) filters. However, complex modifications to the analog front-end should be avoided as this is only a minor issue that can be easily solved by the equalization filter and the ISSR section in the digital back-end of the receiver.

Looking back to the start of this section, the bandwidth compression technique described here is in fact a very basic implementation of a matched filter receiver. By looking at the problem from a unusual point of view, it was found that the classic heterodyne topology can be transformed into a pulse-based radio system with only minor conceptual changes. The only important thing to remember from this section is that a pulse-based radio can be seen as a modified version of a single-carrier radio system. A considerable portion of the carrier is cut away just before the RF-signal is applied to the antenna. However, the internal workings of the system are no different than for any other QPSK transmitter–receiver configuration.

As a closing note, it is stressed again that the main focus of pulse-based wideband radio should be on the problem that is being addressed. In this case the goal is to improve the multipath resolvability of the system in an indoor channel with considerable multipath scattering but a limited delay spread. An approach that focuses too early on *pulse-shaping*, *pulse modulation* or the *Ultra* in Ultra-Wideband is in danger of getting caught up in low-level details (Figure 5.8) and may end up in a deadlock situation.

5.2 Synchronization

The previous section has introduced the basic ideas behind pulse-based wideband radio. It was shown that the names 'pulse-based radio' or UWB may be a proper description of the characteristics of a pulse-based radio signal in respectively the time and the frequency domain, but are in fact very misleading names when it comes to understanding the larger picture. After all, it is quite

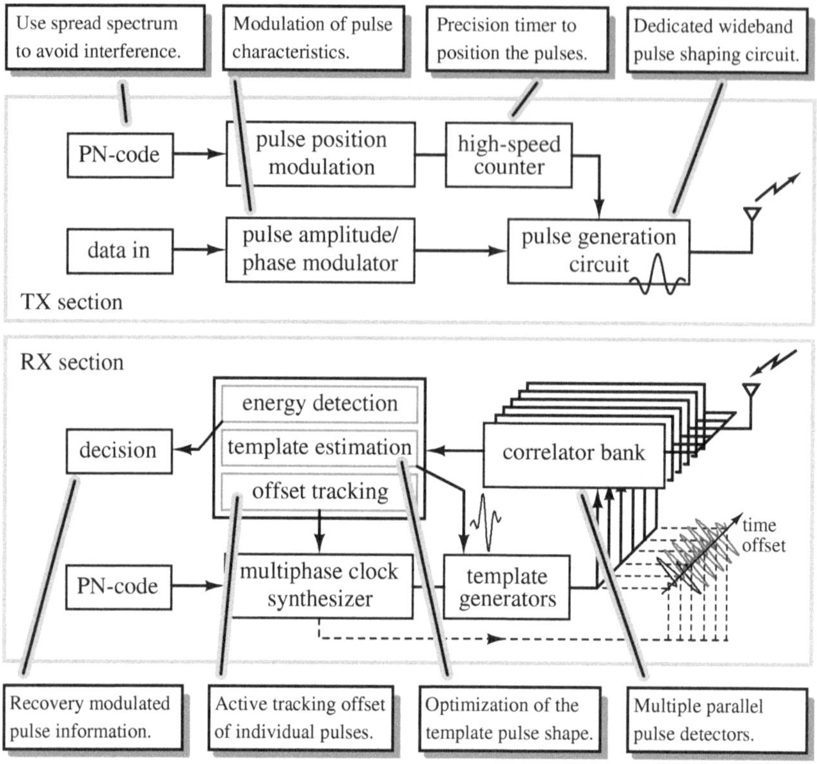

Figure 5.8. How not to build a pulse-based radio. Many pulse-based systems in the literature [Kim08, Ded07, Lee04, O'D05] focus on the generation and detection of pulses. However, this approach tends to get caught up in insignificant low-level details and secondary system patches.

pointless to put endless effort in the realization of a pulse-generating circuit (Figure 5.8) that perfectly fits into the spectral mask imposed by the FCC, then only to forget about the original purpose or ignore new problems that emerge from this approach.

One of the most challenging problems for the designer of a pulse-based radio receiver is synchronization, i.e. trying to lock on a stream of pulses (Figure 5.9). If the receive window is not aligned properly with the arrival time of the pulses, the receiver is unable to 'see' the pulses. This issue is caused by probably the most significant difference between time- and frequency-domain based filtering techniques: a frequency domain filter does not require an absolute phase reference in order to discriminate the signal-of-interest from all other unwanted frequency components. This in contrast to a time domain filtering method, which requires an absolute time reference. The window at

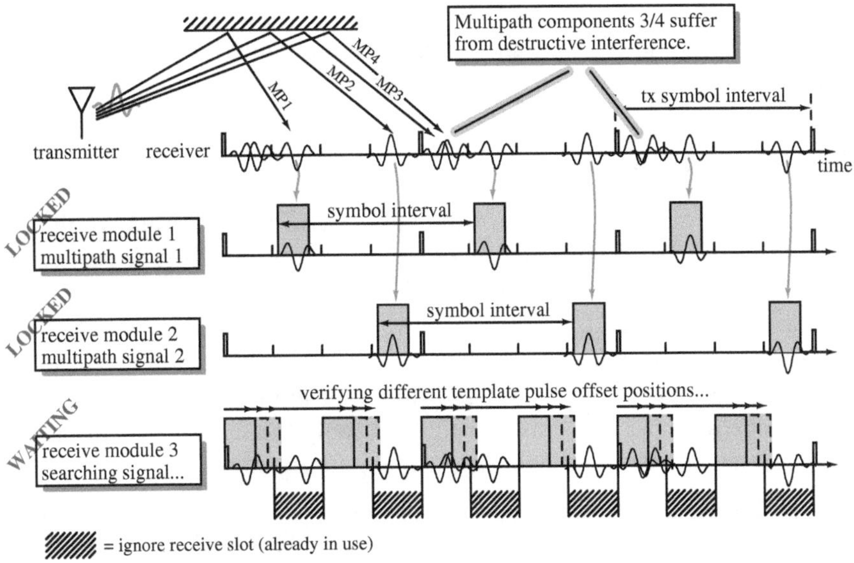

Figure 5.9. During synchronization, a pulse-oriented radio can perform an actively controlled search for the best performing correlation slots over the interval between two pulses. This requires a lot of synchronization time overhead, while the result is valid for only a short period of time in a fast varying channel.

the input of the receiver can be considered as a genuine filter. Signal components (e.g. interference) that arrive during the off-state of the window are effectively blocked. However, a small alignment error of the receive window also eliminates the wanted signal components.

Even after a successful lock has been achieved, the receiver of a pulse-based radio system should be able to track the clock offset of the transmitter with respect to its own time reference. Failing to do this will result in the pulses shifting outside the receive window which again results in signal loss. Several techniques have been proposed to solve the tracking problem. For example, in the early-late delay-locked loop (DLL) approach [Foc01], the receiver attempts to detect clock skew between the transmitter and the internal clock reference by performing three correlations on each received pulse. One correlator is ahead in time with respect to the reference correlator, while the other is shifted slightly later. If the receiver is not perfectly aligned to the stream of arriving pulses, the power emerging from the early and late detectors becomes unbalanced which allows the receiver to take the appropriate countermeasures. Remark that this technique is in fact a very primitive form of the I/Q phase tracking loop used by coherent receiver architectures. The early-late synchronization technique is also a prime example of how to completely make a mess of things. Of course, the early-late detector will probably do the job for which it's designed. But it is a rather quick and dirty fix and ignores the real underlying problem that is explained below.

The power of statistics

Remember the original purpose of the pulse-based system: improve the reliability of a wireless link by increasing multipath resolvability. A pulse-based radio system can achieve this goal by using a setup of parallel receivers, each monitoring a distinct and resolvable multipath signal. It is clear that the maximum number of channels a pulse-based receiver is able to monitor in parallel is practically limited in terms of cost, chip area and power consumption. For performance reasons, it is thus essential that only the multipath streams with the best signal quality are monitored, while ignoring all other components.

So much for the theory. In practice, the receiver has no knowledge of where to start to look for the strongest multipath stream and must rely on an exhaustive search procedure. This means that, during the synchronization sequence in the beginning of each transmission burst, the receive window must scan over a time interval specified by the period between two successive transmitted symbols. The larger the ratio between the pulse period and the window length, the longer it takes before a successful sync can be achieved. Each possible window position that has to be examined increases the duration of the synchronization procedure, generating a significant amount of overhead. Of course, the system can allocate all available receiver modules and perform a parallel scan over the search space, but this is at the cost of extra power consumption and is thus not always a viable option for an idling receiver which only monitors the channel for incoming messages. Another counter-argument against the synchronization procedure described above is that the search results become useless in time-varying channels with a short coherence time: by the time the synchronization sequence ends and the payload of the received packet arrives, the window position for the best multipath component is already based on outdated channel measurements. Note the considerable complexity of the synchronization procedure, while tracking the clock offset between transmitter and receiver is not even brought into play yet.

Again, the reason for this is a wrong approach of the problem, caused by too much attention for the 'pulse'-thing. A more reliable way to deal with the unreliable nature of the channel is to use the unpredictability against itself. By relying on the power of statistics, as briefly explained in Section 4.4. The solution to the synchronization problem requires a small, but essential modification to the pulse-based coherent receiver concept that has been outlined in the previous section: instead of using equal window lengths in both transmitter and

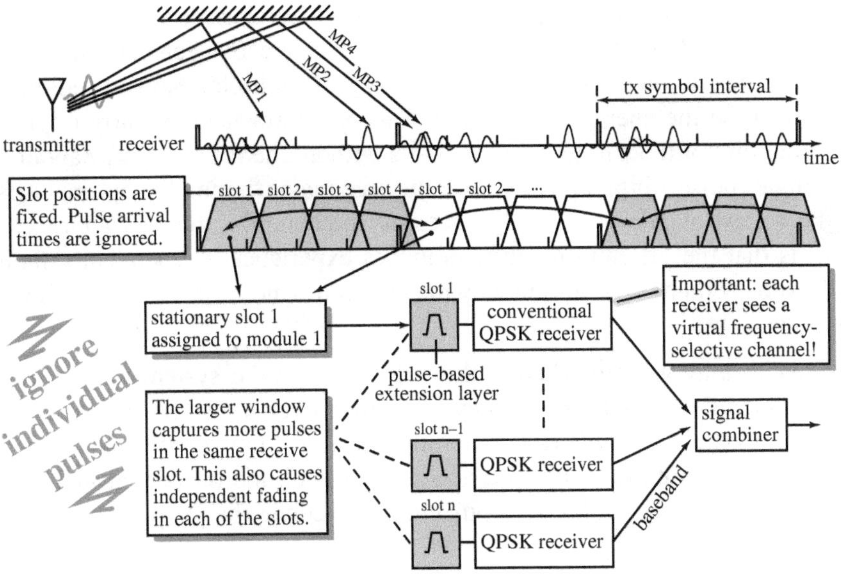

Figure 5.10. In the proposed pulse-based receiver, the interval between two transmitted pulses is divided in a number of fixed receive slots. Each of the receive slots is assigned to a conventional receiver with pulse-based extension layer. The fact that pulses are being processed can be completely ignored in this approach.

receiver, the duration of the receive window is increased. The aim is to cover the entire period between two successive received pulses using the joint effort of all receive modules in parallel. Doing this way a fast single-shot synchronization is always guaranteed (see Figure 5.10). For example, suppose that the interval between two pulses ($T_p = 500\,ps$) is $T_s = 10\,ns$, while the system accommodates 10 independent receive branches. It follows that a minimum window length of 1 ns is sufficient to cover the entire 10 ns time interval at once.

Of course, increasing the receive window length has some inescapable side effects. Increasing the window length in a heavy multipath (indoor) channel increases the probability that two or more multipath-delayed versions of the same pulse pass through the same receive window. This is due to a reduced multipath resolvability of a receiver with an increased window length. The effect is comparable to intra-symbol interference in the classic continuous-time modulated systems, and causes a time-varying alternating pattern of constructive and destructive interference in the affected branch of the receiver. In the proposed scheme, the receiver does not show any intention to avoid this from happening. Both the offset and the duration of the receive window remain untouched.

Instead of this, the receiver passively relies on a combination of statistics and the delay spread in the channel. This strategy has already been extensively discussed in Section 4.4: combining the energy of multiple resolved multipaths – in this case the energy of the parallel receive branches – greatly improves the reliability of the wireless link. All this without the need to worry about every single 'pulse'-thing that arrives at the antenna or finding the optimal offset of the receive window. The only necessary precondition for this approach to work is that the different receiver branches experience independently fading multipaths, which is the reason why the transmit window should be as short as possible.

The reader may wonder whether the pulse-based radio system with multiple receive modules actually offers any advantage over the OFDM-based system with a comparable sample rate. After all, except for the fact that data is being processed by multiple receive branches in parallel, the overall throughput of raw data symbols is not less in a pulse-based receiver than it is for the OFDM-system.

Parallel processing

There are many benefits of parallel processing, though. A parallel architecture reduces the power consumption while maintaining the same throughput as the single-path OFDM system. Also, the system of parallel receive modules allows to save additional power by disabling some of the branches in case of a sufficient signal strength in the remaining paths. Depending on the actual performance requirements, the pulse-based wideband radio system allows flexible scaling, without making changes to the specifications of the transmitted signal: the designer of a cheap, battery-powered system could decide to reduce the multipath resolvability of the receiver in exchange for a reduced level of parallelism. In addition, probably the most underestimated energy saver of the pulse-based radio system is that in-band interferer power is effectively divided over the parallel paths, resulting in a drastic reduction of the linearity requirements (IIP3) in each of the front-end modules in the receiver. For example, if the on-time of the receive window is one tenth of the total period between successive pulses, in-band interferer power is suppressed by 10 dB. In contrast, the front-end of the OFDM-based system is exposed to the full interferer power. The parallelism of the pulse-based radio architecture also allows the use of more sophisticated algorithms to improve the performance of the receiver, even without the cost of extra hardware. Refer back to the example above, where the interval between two pulses is uniformly divided over 10 receive units. Instead of trying to cover the complete interval, it is also possible to eliminate half of the parallel units using an alternating pattern, as shown in Figure 5.11. If the delay spread of the channel is larger than the duration of two receive windows,

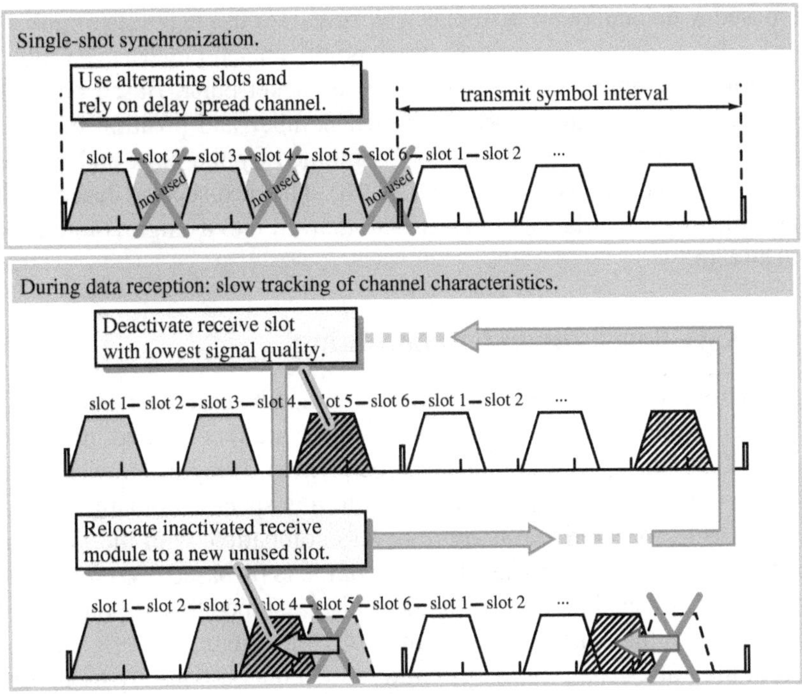

Figure 5.11. During synchronization, all available receive modules are evenly distributed over the interval between two consecutive pulses. During data reception, the receiver tries to improve its performance by relocating the slot with the lowest signal output to the oldest unused time slot.

it can be guaranteed that during synchronization sufficient signal power shall be captured by the remaining receive units, still allowing the receiver to lock on an incoming transmission burst.

As soon as synchronization has been achieved, the system can start to enhance its performance using the following algorithm (illustrated in Figure 5.11). First, the receive branch with the poorest signal quality is selected. Based on an internal history table which contains the average signal quality for each window position, the window of the selected branch is relocated to a new position which is not in use at that moment. During this relocation process, the wireless link remains fully operational thanks to the $n-1$ receive units that are still online. After the disconnected receiver is brought up and is running again, it is hot-plugged into the system after which the whole procedure starts over. This way, the receiver can adapt to slow variations in the average characteristics of the channel without having to chase after every individual multipath component. It is emphasized again that the above procedure is entirely client-based, without any intervention from transmission side. As a result of this feature, the pulse-based wideband radio system is well suited for use in point-to-multipoint networks. For example, a wireless access point can send interleaved information destined for multiple clients in a single packet burst. Based on channel observations, clients can choose their own number and position of window slots. The signal energy of the slots is combined by the back-end processor which controls the receive units (Figure 5.10). The next section describes how the multiple incoming data streams are interleaved into a single, more reliable output stream.

5.3 ISSR-based diversity combining

In this section, it is explained how several multipath streams originating from the parallel units of a pulse-based radio receiver are recombined into a single stream. Even though the system is described with a strong bias towards pulse-based radio, its applicability is not limited to this specific domain. Any other wired or wireless system that is more or less compatible with the ISSR signal reconstruction system can make advantage of the ISSR-based diversity combining method introduced in this section. It is also assumed that the output signal originating from the parallel receive units is a single-carrier QPSK modulated signal. In the pulse-based radio system of which the outlines are being drawn up in the course of this chapter, the ISSR-based stream combiner should be localized in the digital back-end of the receiver. This implies that some crucial signal-preprocessing steps have already been performed autonomously by the respective receive units themselves. This includes, for example, the maximization of the loading factor of the AD-converters using a variable gain amplifier (VGA) and the removal of out-of-band signal components by fixed-frequency band-select filters. Also, the conversion from RF to baseband – in this particular case the bandwidth compression of the RF-pulses (p. 93) – is supposed to be taken care of by the receive units.

Multi-antenna spatial diversity

With the combination technique that is going to be introduced below, it is not only possible to interlace the multipath symbol streams that are resolved from a single antenna, but also from two (or more) physically separated antennas.[15] At first sight, it could seem a bit pointless to introduce costly hardware in the form of additional antennas. After all, remember that one of the motivations behind the development of pulse-based wideband radio was the excellent multipath resolvability of this system thanks to the use of short pulses. So there is no apparent advantage in the reallocation of some of the receive units to another

[15] Other possibilities are separate wires, virtual antenna beams from a phased-array grid, ...

antenna. There is a good reason why this matter is brought to the attention of the reader, though. Wideband antennas are very difficult to design. Matching problems over a wide spectral range (500 MHz+ for the UWB specifications by the FCC) and the physical dimensions of the antenna causes dispersion of the signal. Even for a perfectly matched antenna, any conducting element in the neighborhood changes its impedance. The result is that the RF-energy of the pulses is spread over a longer time. The time-limited pulse that is transmitted, collapses during transport and different frequency components will arrive with different time delays at the output of the receive antenna.

For the receiver, it makes no sense to make the window length shorter than the total dispersion time caused by the transmit or receive antenna: dispersion invalidates the assumption of independent fading between the distinct parallel receive slots, which is necessary precondition to increase reliability. The reader should be aware that is not an ISI-related problem, since it only affects the multipath resolvability of the pulse-based radio receiver. The best way to compensate for the reduced number of slots per receive antenna is to increase the number of physical antennas.[16]

Architecture of the interleaved ISSR decoder

The purpose of the interleaved ISSR decoder described here is to merge multiple independently fading symbol streams which are captured by parallel receive units into a single output stream with a more reliable signal quality (i.e. a lower standard deviation on the received signal power). The entire process of signal recombination occurs before symbol-to-bit demapping, which means that no permanent or irreversible decisions about the received symbol are taken by the interleaved ISSR decoder itself. This task is entrusted to the error correction mechanism which is located further on in the signal processing chain. For more information on the topic of ISSR, the reader is referred to the ISSR decoder introduction in Section 5.3. Remember that the internal workings behind the scenes of ISSR are based on the reconstruction of frequency domain information which has gone missing as a result of frequency-selective fading. As a result of intersymbol interference, each of the parallel receive modules of the pulse-based radio receiver offers a frequency-selective fading symbol stream. It would be an interesting approach if the available frequency bands from every symbol stream could be joined (Figure 5.12), in the attempt to fill in missing holes in the frequency spectrum as much as possible. Any parts of the spectrum that remain missing, or are unavailable in all streams due to narrowband interference are subsequently reconstructed using ISSR. It is evident

[16]Note that the physical distance between the antennas must be large enough to ensure independent fading characteristics ([Fuj01] p. 183).

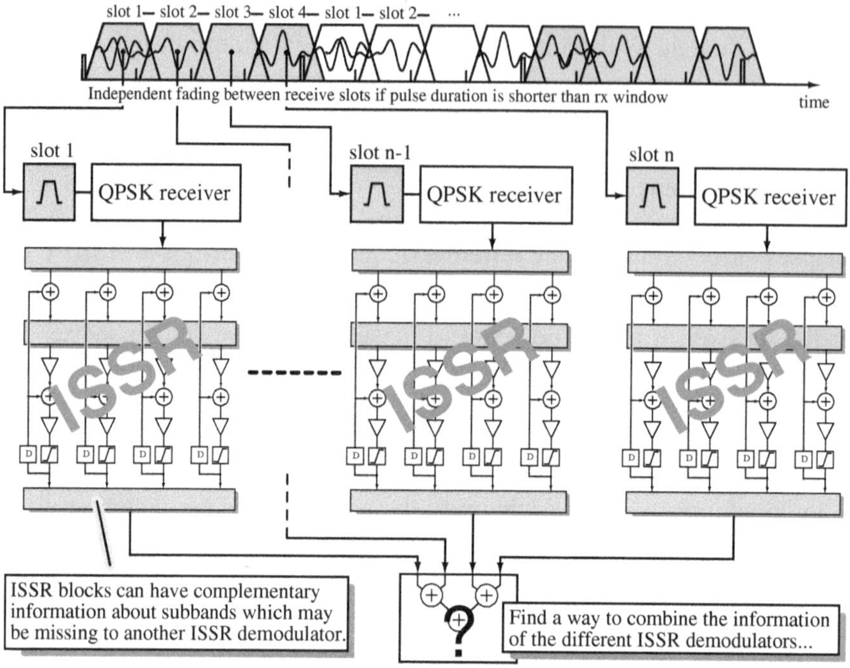

Figure 5.12. The individual data streams captured from the different receive slots suffer from independent frequency-selective fading as a result of ISI. Interleaving multiple ISSR decoders could fill in the gaps of missing frequency bands and leads to a more than proportional increase in performance.

that the more pieces are added to the puzzle, the better chances are that the original transmitted stream can be successfully reconstructed.

The architectural outline of the interleaved ISSR decoder is obtained from the original ISSR decoder as follows. First, suppose that every receive unit in the pulse-based radio system has its own dedicated ISSR decoder module, as is being illustrated by Figure 5.12. Of course, when all ISSR modules operate independently from each other, there is no cooperation that might lead to a more-than-proportional increase of productivity. For example, it is perfectly possible that two receive branches experience complementary fading channels, but no sufficient information is available to either of the ISSR modules to achieve significant results. The next step is to force the ISSR modules to share their information. The interconnection network between the modules is shown in Figure 5.13. It can be seen that the interleaved ISSR decoder shares the entire signal path, starting from the multi-input integrator section. This way, multiple receive units feed their information into the interleaved ISSR decoder section,

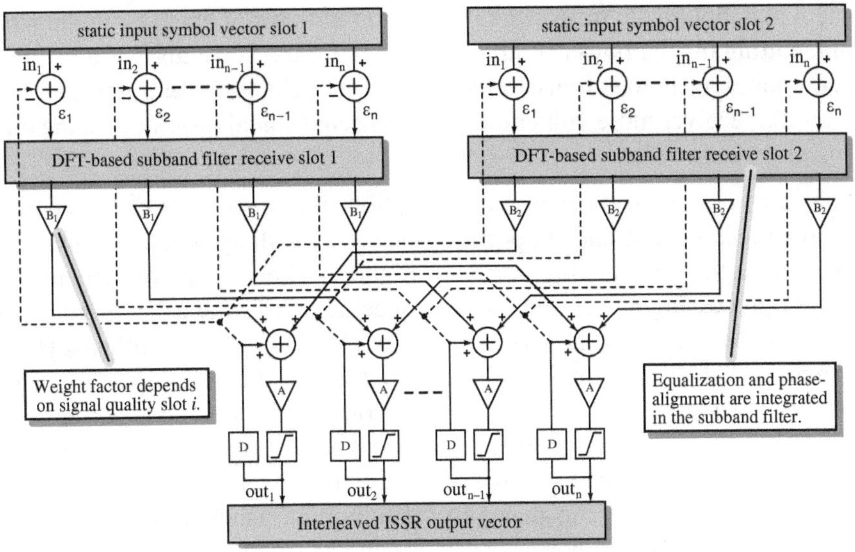

Figure 5.13. Simplified architecture of the interleaved ISSR decoder, for the case of two receive slots. The interleaved ISSR decoder can be easily extended to an arbitrary number of input slots. The weight of each receive slot i is adjusted by gain factors B_i.

and the reconstructed signal becomes available at the output in the form of a single vector. The output signal is then fed back to the individual non-shared sections of the ISSR module, where the relevant portion of the baseband spectrum is compared with the incoming symbol vector from each of the receive units. Before the result of this comparison is reinjected into the common part of the ISSR decoder, the FFT-based filter corresponding to the respective receive unit removes frequency bands that do not contain valid information. This process is repeated for every branch separately, based on continuous signal and interference monitoring measurements. The circle is completed by merging the time-domain error vector of every receive branch in the multi-input integrator block of the shared ISSR section.

Asymmetrical processing branches

The previous analysis was done under the assumption of a flat power distribution over all receive slots in the period of time between to pulses. In a line-of-sight (LOS) channel, where the direct path between transmitter and receiver offers a much stronger signal than the indirect propagation paths, the equally distributed power assumption is no longer valid. For this reason, allocating equal resources to each of the parallel receive branches in a pulse-based radio system may not be the most efficient solution in terms of energy. On the other hand, cutting off the power of all except one of the receive units is a bad idea since it takes away the ability to monitor varying channel conditions and also makes the receiver more vulnerable to unexpected signal drop-outs in the only remaining receive slot.[17]

In order to cut down on power consumption, the receiver is equipped with one main receive unit and supported by several auxiliary units. The main receiver has a superior analog signal path and is allocated more resources in terms of processing power and linearity specifications. The exact calculations or simulations are out of the scope of this text, but in practice significant power savings are achieved by reducing the resolution of the AD-converters of the auxiliary receivers. In many cases, a reduction of the bit depth is not only advantageous in terms of chip area, but this also reduces the capacitive load of the converter on the analog signal chain. Less supply power is thus required in the analog output drivers to maintain the minimum bandwidth requirements. However, the savings on power consumption can go much further than some low-level system modifications: in many circumstances, the quality of the line-of-sight path is sufficient to ensure a correct reception of the signal. It is thus not necessary to feed all information from the auxiliary receive branches into the interleaved ISSR decoder. A considerable amount of processing cycles can be saved by keeping this information on standby. This implies that the data from the auxiliary receive units is temporarily stored in a set of sliding window buffers and that the channel conditions of the associated receive slots are updated on a regular basis, but no further interpretation is done. When the interleaved ISSR decoder detects an imminent problem of signal quality in certain frequency bands, it can decide to call in the help of the auxiliary branches. The previously stored information from the relevant branches is then injected into the running signal reconstruction process and some extra ISSR cycles are scheduled to allow the system to converge. The quality of the signal can be monitored using the error vector of the ISSR loop.

If the error vector contains an unexpectedly high portion of unavailable frequency bands, this may point to the fact that insufficient information is available for the ISSR loop to reconstruct the signal vector and extra receive branches must be activated. An important distinction must be made between the case where the energy of the error vector is concentrated in a number of bands or the case where the problem is spread over the entire baseband spectrum. The latter observation indicates that the problem is possibly caused by background noise in the channel, rather than it is the result of ISI. There is little the ISSR decoder can do in the latter situation, so the recovery task should be handed over to the error correction subsystem.

[17]...which is in fact a reliability problem caused by a decreased degree of multipath diversity.

5.4 System integration and clock planning

As a result of the very specific technical knowledge and the development costs associated with in-house IC-design, end product vendors are shipping devices that contain proprietary IP blocks from several different sources. In the never-ending quest for cheaper and more reliable products, the vendors of IP components try to offer off-the-shelf system-on-chip (SoC) implementations. In many occasions, offering single-chip solutions is the only way to differentiate themselves from their competitors. For wireless products this can be a major challenge, since the analog circuits have to share the same die as the digital signal processing circuits. For the inherently very sensitive analog circuits of a wireless receiver this is a real performance killer, as their is possibly no better way to get a broadband noise source any closer to the analog front-end.

There are several options for the digital section to annoy its analog neighbour to death (illustrated in Figure 5.14). For a start, it can use the shared supply lines to get the job done. If some smart designer would have taken this into account and has avoided common power lines, the digital part will be smart enough to deliberately make use of the antenna functionality of the bonding wires and get the message delivered over the air. If this also fails due to a clever floorplanning

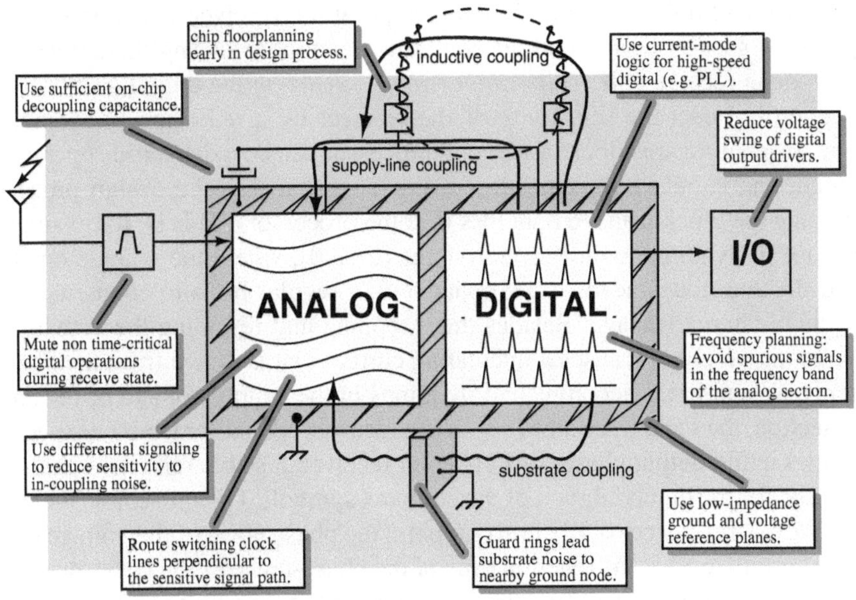

Figure 5.14. Ways to reduce noise injection into the analog section of a single-chip receiver design with on-board signal processing. Injection of digital noise in the front-end of the receiver reduces its sensitivity to the weak antenna signals.

of the chip integration team, the digital section can revert to the last resort and go underground: through substrate coupling and a finite common mode rejection ratio (CMRR), switching noise will find its way to the most sensitive nodes of analog signal path and decreases the sensitivity of the receiver. Several techniques exist to shield the noisy digital circuits from the analog section: on-chip decoupling of power supply lines or guard rings that pick up unwanted substrate noise and lead it away to a stable ground connection. Also, the analog circuits themselves can be hardened against external in-coupling noise by an extensive use of differential signaling.

Other approaches go directly to the source of the problem by twiddling with the frequency at which the digital part is clocked. This way spurious emissions at some higher harmonic of the clock signal can be kept away from the frequency band-of-interest. Sometimes, clock-dithering is also being used in switching circuits to avoid single-tone spurious emissions. However, dithering merely spreads the same amount of noise power over a larger chunk of the radio spectrum and is thus of very little interest for wideband receivers. For the pulse-based radio receiver, there is still another way to prevent noise from leaking into the front-end of the system. The input of the receiver is most vulnerable to noise during the on-state of the receive window. For the remaining time, the input stage is kept in some sort of low-impedance state where interference and noise are blocked from entering the receiver. It is possible for the digital section to suspend all non-critical activities during the active slot of the main receive unit. This form of *radio silence* in the digital section will temporarily boost the sensitivity of the receiver by a reduction of the noise floor. Suspending the clock of a digital processor can be a dangerous operation though, and should be taken into consideration early in the design process. Stopping the clock leads to data loss in some processors. This is, for example, the case in dynamic logic CMOS systems [Cha00], where the voltage on isolated floating nodes needs to be updated on a regular basis to counteract the effects of charge leakage. Besides this, stopping and resuming the activity of a large digital circuit causes a fluctuating current that is drawn from the power supply. When this is accompanied by jumps in the supply voltage of the analog section, the local oscillator (LO) of the receiver will suffer from unexpected changes in the output phase. In a coherent receiver, a stable reference phase is essential for the demodulation of the complex symbols. Of course, it is the task of the XTAL reference oscillator to suppress the phase noise of the voltage controlled oscillator, but the reaction time of the phase-locked loop (PLL) depends on the reference frequency and the bandwidth of the loop. In order to allow the LO-phase to stabilize, the digital clock should be muted slightly in advance of the start of the next upcoming receive slot. When the RF-signal has been captured and the receive window is shut, the activities of the digital section can be resumed.

Reference phase of the local oscillator

A very important distinction between the analog front-end and the digital back-end of the pulse-based radio system (and in general for any coherent modulation method) is that the local oscillator of the analog front-end in both transmitter and receiver must be kept up and running, even when the receiver is idling in between two consecutive pulses. Putting the RF-oscillator in a sleep state in the urge to save power would be a big mistake, even if the reference crystal oscillator of the PLL remains active. This is because the phase reference of the I/Q signal for the up- or downconversion mixers is ambiguous after the startup of the PLL.

Eliminating the I/Q phase ambiguity is one of the purposes of the training sequence at the start of every transmission burst. The loss of the phase reference makes it impossible for a coherent demodulation system to correctly demap the next complex symbol. Between two consecutive data bursts it is perfectly possible to schedule downtime of the entire analog front-end to save power. But the reader should keep in mind that it is not the primary goal of a low-PSD wideband radio system to offer a power efficient solution for low-throughput applications.

During the circuit-level design of the pulse-based radio receiver, which will be discussed in detail in Section 5.2.6, several low-level design techniques have been used to suppress the noise from switching logic circuitry as much as possible.

For example, circuits that operate within the input frequency band of the receiver use current mode logic (CML). Compared to conventional CMOS logic, current mode logic uses analog design techniques to represent digital signals: CML circuits represent digital information in a differential manner and draw a constant current from the supply lines. Doing this prevents that switching noise is injected into the power supply. While for low frequencies current mode logic is commonly avoided because of the direct current path to ground, this particular disadvantage largely disappears for high speed signals, such as those encountered in the prescaler of a PLL circuit. The reason is that for high frequencies, the dynamic power consumption of standard CMOS logic becomes comparable to the static power consumption of CML. The same technique has been used to drive off-chip data and clock signaling lines. A lot of instantaneous current is involved in the charging and discharging of the parasitic capacitance of external components and it is not a good idea to use the supply lines as the return path for those peaking currents. The receiver implementation which is described in the next section uses analog differential current drivers with a reduced voltage swing for this purpose, a method that is closely related to LVDS.[18]

A last design detail worth mentioning is that the internal clock distribution lines are oriented perpendicular with respect to the RF-signal carrying path, in order to minimize inductive coupling between both. The distance between the clock buffers driving these lines and the input stages of the receiver must be as large as possible and a common supply voltages should be avoided as much as possible. At the same time, low-impedance ground/supply reference planes are indispensable in many cases, since the clock lines are not terminated differentially but directly drive (charge/discharge) the gate of CMOS transistors. The return path of the clock signal is thus formed by the closest AC-ground terminal (i.e. the ground- or power plane) instead of the inverse clock line in a differentially clocked system.

5.5 Comprehensive overview of the pulse-based radio system

This section provides a more general overview of the pulse-based radio system and briefly explains the coming about of its underlying principles. The whole story starts with the FCC deciding to open up 7,500 MHz of spectrum in the 3.1 to 10.6 GHz frequency band for use by unlicensed radio communication systems [Com02]. The considerable amount of bandwidth involved here offers the potential for new possibilities in terms of wideband communication networks and immediately draws the attention of both industry and academics. Ultra-Wideband (UWB) communication systems as they are called, became soon marketed as being the wireless cable replacement technology of the future. What is often not well recognized is that the maximum power spectral density (PSD) which these wideband communication devices are allowed to emit is very low: less than −41.3 dBm/MHz EIRP.[19] In fact, exactly the same limits were already into force for a long time for unintentional radiators (computers, switching power supplies). The main difference that there is at this moment no further discrimination based on the origins of the transmission (i.e. intended or not). It follows that there are some inevitable drawbacks that need our attention.

[18] LVDS: low voltage differential signaling [Ass01].

[19] EIRP: Effective Isotropic Radiated Power.

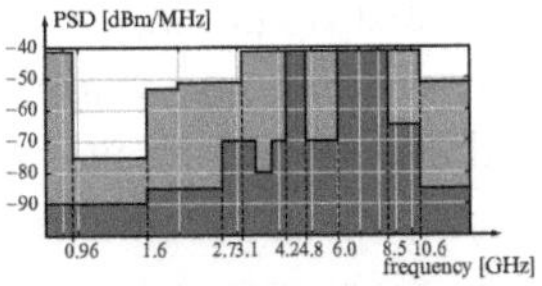

The total power that can be transmitted in compliance with the Part 15 rules imposed by the FCC is remarkably low: even when integrated over the entire frequency range of 7.5 GHz, the maximum output power is in the order of half a milliwatt (−2.5 dBm). Compare this to the power levels that are employed by the 802.11a system (+23 dBm and higher), and it should be clear that those two orders of magnitude must have some repercussions regarding the signal-to-noise ratio of the received signal. Also, as exemplified by the 802.11a system transmitting in the middle of the UWB spectrum, there has never been any intention to allocate this very large frequency band exclusively for low-PSD wideband radio. The tolerance against in-band interference is a difficult design issue and at the same time the most important performance measure of a wideband radio system. Maybe the whole issue is best described by the label that is found on many Part 15 compliant devices:

> CFR 47 PART 15.19 LABELING FOR LICENSE-FREE RF-GENERATORS:
>
> Operation is subject to the following two conditions: (1) this device may not cause interference, and (2) this device must accept any interference, including interference that may cause undesired operation of the device.

Disadvantages of OFDM in wideband radio

In other words, a Part 15C intentional transceiver system has no legal rights at all and should find its own way around interference that suddenly turns up in the middle of its frequency spectrum. A wideband radio system should dynamically react on changing channel conditions and must be able to quickly reallocate resources to frequency bands with an acceptable noise and interference level. This is probably the main reason why the attention of the industry has progressively shifted towards a multicarrier approach based on the OFDM system. It is believed that the flexibility of OFDM is the most reliable answer to unstable parameters that come with a wideband channel such as narrowband interference (NBI) and intersymbol interference (ISI). OFDM – originally developed for high-PSD communication systems – is a system that entirely relies on digital signal processing, and while it is the most flexible system from a system engineer's point of view, it entirely relies on a digital representation of the baseband signal. For a radio system with a large spectral footprint, this implies that the entire signal chain from antenna down to the digital back-end must support this large bandwidth. Of course, with the obvious consequences related to power consumption. Unfortunately, the increased efforts are not necessarily reflected in a decent data throughput in an Ultra-Wideband system. The low spectral density of the received signal forces the designer to use a limited modulation depth (BPSK/QPSK) in the OFDM subbands. But reducing the modulation depth does not result in more relaxed linearity requirements in the front-end of the receiver, because accidental in-band interferers are only filtered later on in the digital back-end.

Based on the aforementioned observations, it seems that OFDM may not be the best option for wideband radio communication after all. The underlying reason for these problems is that the real opportunities of wideband radio are being ignored. First, one must realize that no earth-shocking results are to be expected from a system that has a very large spectral footprint, but is only allowed to use a very limited amount of transmission power. So what is the actual advantage of using such a large spectral footprint? The answer being looked for is found in the characteristics of the propagation channel. A wireless communication device that is operated in an indoor environment suffers from a very specific form of multipath fading. The delay spread (τ_{RMS}) of the channel is limited to a few hundreds of nanoseconds, which at first sight may sound very promising because of the limited amount of intersymbol interference. However, this also implies that the multipath energy related to a single transmitted symbol arrives in a very concise time frame at the antenna of the receiver. A narrowband receiver will see a flat fading channel, because the delay spread of the indoor channel is below its resolution bandwidth.

Increasing the resolution bandwidth as a means to improve diversity

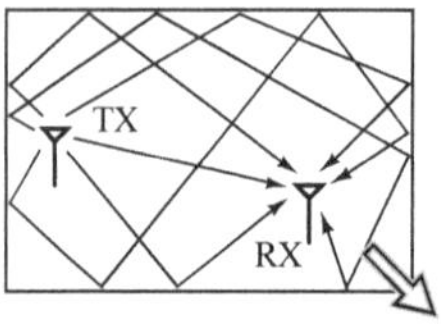

A wireless system cannot make a distinction between multipath components that are below its resolution bandwidth. From a frequency domain point of view, a channel with a low delay spread corresponds with a frequency-selective fading channel, while flat fading is experienced over wide portions of the frequency spectrum. A transmission with a spectral footprint smaller than the coherence bandwidth of the channel will thus suffer from flat fading, meaning that destructive interference causes intermittent link outages. The designer of a system can tackle this issue in a number of ways. For example, a combination of error coding and using the diversity of time allows the system to bridge short periods of destructive interference. However, this approach turns out not to be very reliable in static channels: the long coherence time here causes long interruptions of the radio link. A second option is to use the spatial diversity of a multi-antenna setup. For a sufficient distance between the individual antennas, each branch of this hardware RAKE receiver system sees an independently fading channel. Switching between or combining the energy from both antennas reduces the possibility that the entire signal suddenly vanishes in thin air. A third way to defend against a frequency-flat fading channel is to raise the spectral footprint of the transmission well above the coherence bandwidth of the indoor channel.

The latter may be the most undervalued advantage of a low-PSD wideband radio system. The frequency-selectivity of the wideband channel is the best insurance that at any moment in time, part of the frequency band is open for transmission. Exactly which frequency channels are available at a particular moment in time is not known at transmission side, due to the time-varying characteristics of the channel. A simple solution to this problem is to spread information in a redundant form over the entire transmission[20] band and let the receiver itself decide from which frequency bands it is most advantageous to extract signal energy. A single-carrier QPSK modulated system is perfectly suited for this purpose, since the information of a single symbol is automatically distributed over the transmission band. In order to suppress the effects of ISI, channel equalization can bring some salvation. In frequency bands where is impossible to achieve an acceptable signal quality due to destructive fading or narrowband interference, the ISSR signal reconstruction technique has been brought into the game.[21]

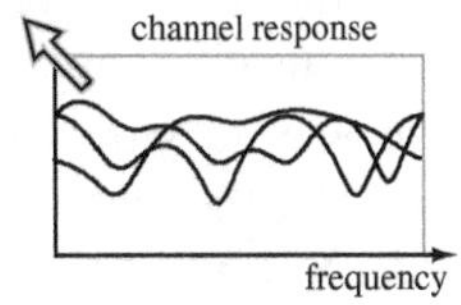

Decoupling symbol rate from multipath resolvability

There is still one hurdle that needs to be overcome, though: the symbol period of the transmission needs to be shorter than the delay spread of the indoor channel in order to make some profit from the diversity of frequency-selective fading. In a continuous-time modulated single-carrier system this would lead to an impractically high symbol rate, with exactly the same problems as those encountered in the OFDM approach: a wideband signal chain and a low power efficiency. In Section 5.1, this observation has lead to the introduction of a system based on short pulses. In such a pulse-based radio, the symbol rate can be decoupled from multipath resolvability (or resolution bandwidth).

The new system is mainly based on the regular coherent QPSK transceiver architecture, on top of which a pulse-based extension layer is placed. While receiver still generates a genuine QPSK modulated signal, only a small fraction

[20] This is exactly what is already being done by the BICM scheme of the OFDM-based 802.11a/g system.

[21] ISSR: Interferer Suppression and Signal Reconstruction, see Section 5.3.

of the RF-signal is effectively *gated* to the antenna. The linearity of the power amplifier is of little importance, because the spectral footprint of the transmission is determined by the duration and the smoothness of the transmission window. The opening and closing times of the transmit window are synchronized with the underlying QPSK transmitter, in the sense that a transition between two constellation points always occurs during the off-state of the transmit gate. The way in which the transition occurs is not of importance, as long as the phase of the carrier has been settled before the start of the next *transmission gate*. Since only a small portion of the modulated carrier is injected into the channel, synchronization of the window with the phase modulator ensures that all necessary phase information is included in the transmitted signal.

The time-domain output of the transmitter has the appearance of a stream of very short pulses, hence the name *pulse-based radio* that is used throughout this text. The period between two consecutive pulses shows small time variations, caused by the phase modulation of the underlying QPSK engine. However, it is strongly advised not to drift away from the original concept and become obsessed by the pulse-thing from this moment on. It is also an unforgivable mistake to try to build a receiver which is based on a 'pulse detector' or – even worse – try to measure the spacing between pulses and call this approach 'pulse position (de)modulation' [Mag01]. Instead of this, the reader is strongly encouraged to consider the transmitted signal of the pulse-based radio system as any other single-carrier phase-modulated system. The only difference being that some non-crucial portion of the time-domain signal is cut away by the transmitter in order to increase the multipath resolvability of the transmission (Figure 5.15). When the stream of uniformly spaced pulses arrives at

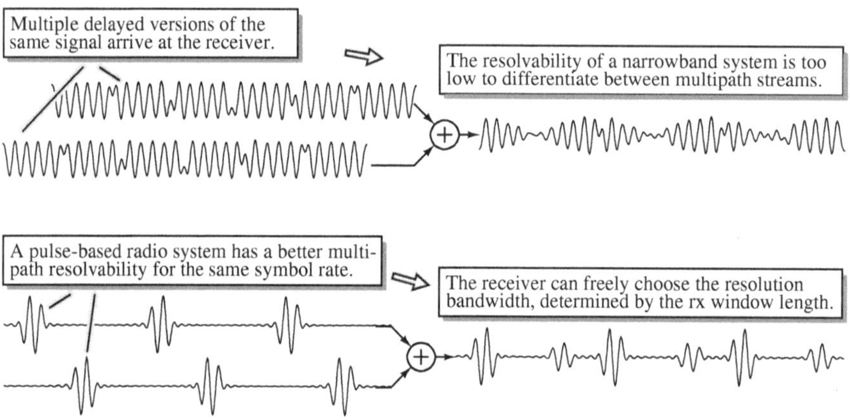

Figure 5.15. Using the same baseband symbol rate, the pulse-based system offers a better multipath resolvability than a continuous-time modulated signal. This approach eliminates the need for an impractically high-speed and power-hungry baseband section.

the receiver antenna, the first task of the receiver is to place a window over the pulses, the same way as is done at transmission side. The main purpose of this *receive window* is to prevent the bulk of unwanted noise and in-band interferer power from entering the front-end of the receiver. In fact, the receive window acts as a coarse time-domain filter and relaxes the linearity requirements of the subsequent signal processing chain. Unfortunately, the benefits of this receive window do not come for free: it is extremely difficult to maintain a correct alignment with the received pulse stream, because the arrival time of the pulses varies due to changes in the propagation path length. All pulse energy that arrives during the off-state of the receive window is lost forever.

Synchronizing on a stream of pulses

To facilitate locking on the pulse stream, the receive window is chosen larger than the transmit window. For example, if the period between two pulses is divided into 10 equal (non-overlapping) receive slots, the search space of the receiver becomes limited to only these 10 slots. The power of unwanted narrowband interference is suppressed by 10 dB, and can be considered as the processing gain of the pulse-based radio system. Increasing the window length has some unintended consequences, though. In a multipath indoor channel, no single pulse stream is being received. Due to multipath, several delayed ghost copies of the same stream arrive at the antenna within a short period of time. On average, an extended window length allows the receiver to capture more pulse energy, which is a good thing. But pulses that arrive within a single receive slot will start to interfere with each other, the result of the reduced multipath resolvability of the extended receive window. There is in fact an equal chance for positive or destructive interference, but this phenomenon results in a reduced reliability (uptime) of the wireless link a static indoor channel.

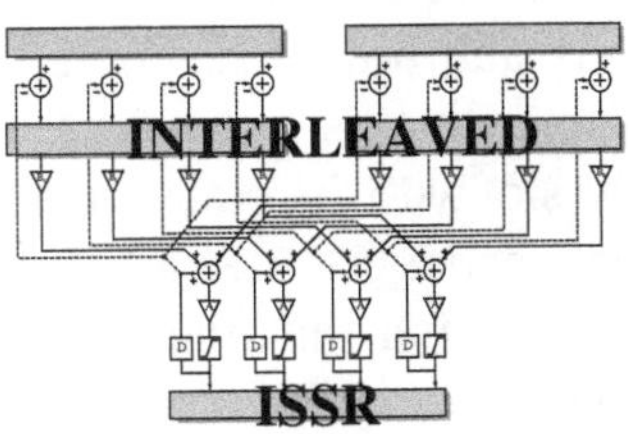

To prevent this from happening, the main receiver is supported by several auxiliary receive units. Each of the auxiliary receivers monitors the channel at a different receive slot, however not every possible receive slot has its own dedicated receive unit. Instead, a receive unit can be dynamically allocated to a slot. The first task of the auxiliary receivers is to help the main receiver to bridge unexpected outages of the main receiver. Data that is captured by the auxiliary receivers is kept in standby and can be immediately injected in the main symbol stream in case an insufficient signal quality has been detected. Each auxiliary unit has its own dedicated equalizer, which aligns the alternate backup stream to the main symbol stream. Finally, the information of the auxiliary receive unit is added to the main signal path in the *interleaved* ISSR decoder, which has been discussed in Section 5.12. Since the duration of a transmitted pulse is shorter than the length of the receive window and multiple pulses can arrive within a single window, each of the parallel monitored receive slots experiences independent fading. It is thanks to the combination of signal power from several independent channels[22] that the reliability of the wireless link will significantly improve.

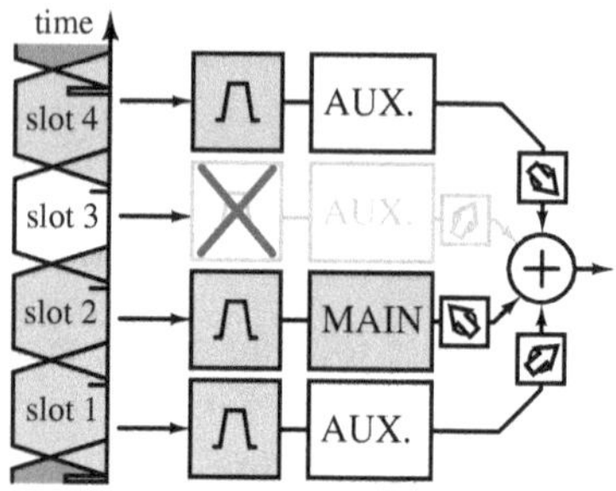

Of course, providing backup information is not the only task of the auxiliary receiver bank. An auxiliary unit also plays an important role in channel monitoring. During the low-power sleep mode of the receiver, only one of the auxiliary units remains awake to monitor the channel for incoming messages. Since there is a reasonable chance that the wakeup call from the transmitter arrives during the off-state of the receive window from the only active receive unit, it can take some time before the receiver becomes alerted. When the transmitter has been able to draw the attention of the receiver, the latter one shall respond with an acknowledgement. At the same time, the receiver also returns to a full power state. The receiver is not synchronized to the transmitter at this moment and it also does not know which receive slots offer the best link conditions. Therefore, all receive units are being assigned to slots that are uniformly spaced over the pulse period. A typical indoor channel will provide a sufficient amount of delay spread (up to a few 100 ns) so that enough multipath power of the synchronization sequence always ends up in at least one activated receive slot.

Immediately thereafter, the receiver reallocates the main unit to the receive slot with the highest signal power. Of course, it is not necessarily the case that all other auxiliary units are being positioned at the best possible time slots. For this reason, the auxiliary unit with the lowest signal-to-noise ratio is disconnected from the active receive segment and is assigned to the slot with the oldest entry in the history table. The auxiliary unit is reactivated when a better SNR value is detected in one of the current inactive receive slots. After this, the whole process is repeats from the beginning, until the end of the transmission burst has been reached.

The algorithm described above allows the system to track average shifts in the power delay profile of the channel, without control track in fast changing environments. Remark that the algorithm is also able to track slow displacements

[22]Each receive slot can be regarded as a separate transmission channel.

of the delay profile that are caused by the clock offset between transmitter and receiver. This causes a slow shift between the mutual alignment of transmit and receive window, but should not be accounted for in the window-generating subsection of the receiver. It means that for the receiver, no fixed relationship is required between the clock of the window-generating circuit and the digital back-end. Clock-offset will indeed cause a rotating constellation in of the received QPSK signal as seen by the signal processor, but this issue can be solved without the need for a feedback path all the way up to the reference local oscillator of the analog section.

Power efficiency and bandwidth compression

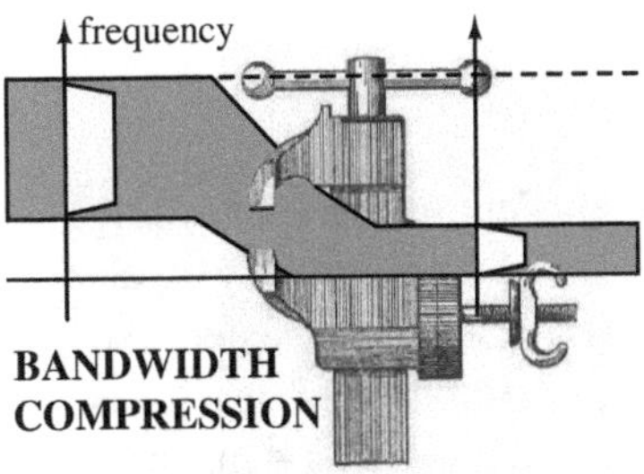

There is still one aspect of the pulse-based radio system that has not been touched on in this overview. It was already mentioned at the beginning of this section that by adding a gated transmit layer to the system, the symbol rate could be decoupled from the multipath resolvability of the system. However, the large bandwidth which is needed to represent such short pulses consumes a lot of power in the analog signal chain of the receiver. For reasons of power efficiency, the receiver must convert the pulse stream to the original continuous-time QPSK modulated signal as soon as possible. From then on, the well-known coherent receiver architecture can be mobilized for all further processing of the baseband QPSK signal. Fortunately, the description of the transition process from RF-pulses to a continuous-time QPSK modulated signal is more complex than the actual implementation. Just like any other narrowband architecture, the RF-pulses are first converted from passband to baseband by I/Q-mixing the received pulse stream with the center frequency of the transmit carrier. At the output of the mixer, of course, nothing has been gained yet with respect to the signal bandwidth. The actual bandwidth compression is obtained by low-pass filtering the I/Q mixer outputs beyond the symbol rate of the transmission. Due to this bandwidth limitation, the signal at the output of the low-pass filter is unable to track the fast changes of the down-converted pulses, and only the envelope of the signal is retained. The I/Q outputs of the envelope detector[23] yield the original continuous-time QPSK modulated symbol stream. From this

[23] Remark that the coherent envelope detector does not suffer from the dead-zone sensitivity problems as the non-coherent diode-detector. In the AM broadcasting service, the dead-zone of diode-detector based radios is circumvented by limiting the modulation depth to about 85% of the peak carrier amplitude. At receiver side, the strength of the residual carrier in the transmitted signal is commonly used to drive the AGC circuit. However, the RF carrier contains no information and is for the transmitter a pure waste of energy.

point on, the remaining part of the baseband receive chain is no different than this used in a regular narrowband QPSK radio receiver with a digital signal processor in the back-end.

In essence, the reader should consider the pulse-to-baseband conversion process as being the equivalent of the passband-to-baseband conversion in a heterodyne receiver. In combination with the coarse interference suppression of the receive window, this type of *bandwidth compression* forms one of the biggest advantages of the pulse-oriented system over an OFDM-based approach of low-PSD wideband radio. Since power consumption has been introduced early in the design process, the parallel processing topology of pulse-based radio system is – by design – far more energy-efficient than the OFDM topology.

LOW-PASS FILTERING FROM AN ENERGY POINT OF VIEW

From a mathematical signal processing (DSP) point of view, rejecting all signal components beyond the symbol sampling frequency is almost a criminal act, since a lot of valuable signal power is wasted. An analog designer sees things somewhat different though. A low-pass filter is typically implemented as a resistor in series with a capacitor. In this particular case, the resistor is determined by the real part of the output impedance of the downconversion mixer (see Figure 6.5). The low-pass characteristic is obtained by adding a capacitor in parallel with the output node of the mixer circuit. The capacitor is charged and discharged by the output current of the mixer, but the only thing that is important for the SNR of the signal is the amount of *signal energy* that gets stored on this capacitor. No matter what the value of this capacitor may be, half of the signal energy is always lost during the transfer from the mixer to the capacitor [Tse95]. This implies that the size of the capacitor – and thus also the location of the pole of the filter – is of no importance whatsoever for the loss of signal energy. It is demonstrated that caution is needed in the interpretation of mathematical results without taking the elementary principles of electronics into account.

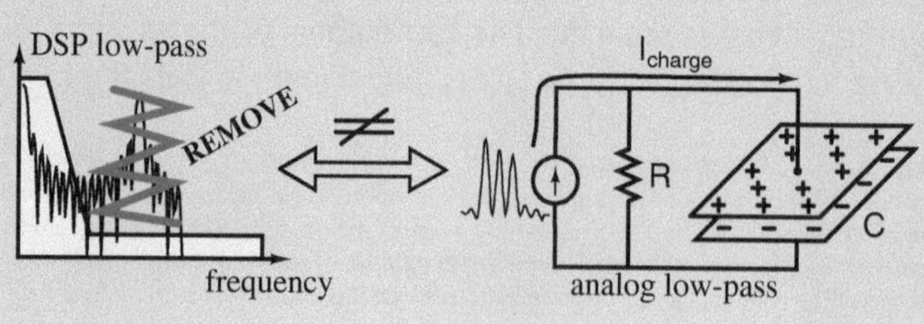

Standardization and backward compatibility

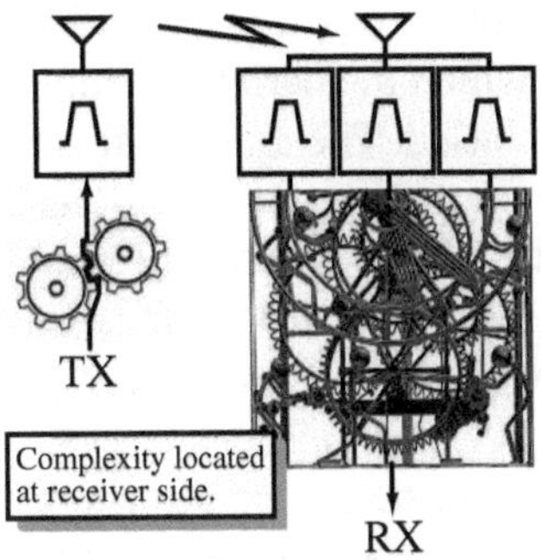

Most of the complexity of the pulse-based radio system is located at receiver side. For a possible standardization procedure, this implies that the actual specifications of the physical layer in the OSI[24] model can be kept at a very basic level, with the synchronization procedure probably being the most complex aspect of the standard. With most of the work pushed to the receiver, the system designer is free to choose the complexity of the implementation, without loosing compatibility with the standard itself. For example, an outdated and basic version of the pulse-based radio receiver with only limited hardware resources will be able to communicate with a highly performing and scalable multi-antenna receiver. Also, the scalability and flexibility of pulse-based radio is much better than this of the OFDM transceiver: based on the actual quality of the signal and properties of the channel, the receiver can decide to reduce the number of active receive units to save power. In fact, the receiver is completely free to change the number and duration of the receive slots, which is something that is completely unique to the pulse-based radio system. This trend is even taken a step further in the digital back-end of the receiver. The processing effort that is put in the ISSR algorithm is entirely up to the discretion of the receiver. If an error is detected later on in the data processing chain, the receiver can schedule some extra cycles of the ISSR algorithm, without interrupting the transmitter for this.

With the same flexibility, the receiver is free to completely unplug the ISSR module from the chain if the signal quality allows for it. And since the transmitter does not play an active role in the reconstruction process, a designer has the freedom to tweak the ISSR algorithm and implement better performing solutions. Also at the hardware level, there are a lot of opportunities to push down the costs of hardware. At receiver side, there is indeed an increased need for chip area which is due to the parallel setup of the pulse-based system, but some building blocks can be reused. For example, the local oscillator and PLL are shared among all receive units (even the transmitter) as they all operate in the same frequency band. The addition of an extra auxiliary receive unit only requires an extra mixer circuit,[25] a quadrature baseband variable gain amplifier and a dedicated analog-to-digital converter.

[24] OSI model: Open Systems Interconnection Basic Reference Model. This seven layer model provides an abstract description of a generalized computer or network protocol architecture [Zim80].

[25] The mixer cannot be shared among different receiver chains for reasons of in-band switching noise.

Conclusion and indicative numbers

As a conclusion of this overview of ISSR-supported pulse-based wideband radio system, some indicative numbers are given, on which the hardware implementation of the receiver in the next section will be based. The main goal is to secure a reliable wireless link over a link distance up to 10 m. Calculations[26] for a channel with only thermal noise predict a theoretical capacity of about 220 Mbit/s. Based on these figures, a QPSK symbol rate of 100 Msymb/s was chosen, which results in a raw data rate of 200 Mbit/s transported over the channel. Note that there is a limitation on the energy spectral density (ESD) of the system, so increasing the symbol rate even more decreases the energy per pulse. A higher symbol rate would also decrease the processing gain of the receive window. At receiver side, the period between two transmitted pulses is divided into 10 equivalent receive slots, which gives the receiver a multipath resolution in the order 1 ns. The in-band interferer suppression ratio of the gated receiver is thus 10 dB.

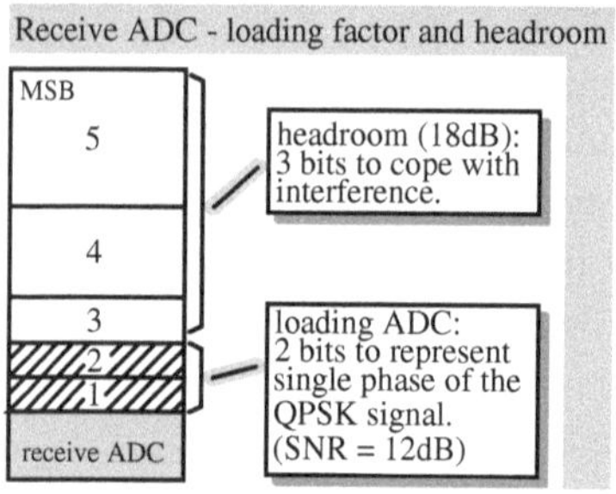

The 3 most-significant bits of the analog-to-digital converter are reserved to cope with the residual in-band interferer power and a suboptimal loading factor of the AD-converter due to the incorrect gain settings of the baseband amplifier. The extra word length of the AD-converter in combination with the suppression factor of the receive window makes that a signal-to-interferer ratio of -18 dB can be tolerated before the sensitivity of the receiver starts to deteriorate, either due to clipping or due to an increase of the quantization noise. If a 5-bit converter is being used (one for the I- and one for the Q-chain), the two lowest significant bits are used to represent the baseband QPSK waveform. This way, the quantization noise is below the thermal noise generated by the channel itself, limiting the implementation loss induced by the AD-converter. The following chapter will discuss the implementation details of the analog signal chain from a single receive unit, starting from the antenna input terminal up to the input of the AD-converter (not included).

[26]Constant-gain, omni-directional antennas with $1/f^2$ aperture, -14.3 dBm transmit power, 5 GHz center frequency and a 1 GHz-wide LOS channel.

Chapter 6

REFERENCE DESIGN OF A PULSE-BASED RECEIVE UNIT

The last section in this book provides a more technical discussion of the implementation details (and problems) of a fully integrated front-end of a pulse-based receive unit. In the full-scale implementation of the system, several of these modules would work together in parallel. They form the backbone of the pulse-based radio system which has been discussed throughout the previous chapters. The wideband receiver front end is intended for use in the 3.1 to 10.6 GHz band which was released by the FCC early in 2002 for use by so-called *ultra-wideband* (UWB) radio devices.

The receiver was implemented in a main-stream 0.18 µm standard CMOS technology without any particular RF-enhancements. Incorporated in the $1.4 \times 1.4\,\mathrm{mm}^2$ chip are (1) a mixed-signal multiphase clock generator which concerts the interactions between different subsystems, (2) a mixer/pulse-to-baseband converter and (3) a baseband amplification chain. The dual in-phase/quadrature (I/Q) signal path of the receiver allows for coherent demodulation of phase modulated pulse-based radio signals. The design is optimized to cope with the large bandwidths at the RF-input stage and the baseband output buffers are able to directly drive an external analog-to-digital converter or measurement equipment. Three selectable sampling speeds are available in the receiver, supporting a maximum symbol rate of 107 Msymb/s. The entire system, which includes the clock drivers and the output buffers, consumes 120 mW from a single 1.8 V power supply.

The block diagram of the pulse-based wideband radio receiver is shown in Figure 6.1. Remark that the architecture is very similar to the signal chain of a (zero-IF) quadrature heterodyne receiver. The most significant distinction with the traditional receiver architecture is formed by the gating circuit located in front of the receive chain. The receive window has two different tasks, each

W. Vereecken and M. Steyaert, *Ultra-Wideband Pulse-based Radio: Reliable Communication over a Wideband Channel*, Analog Circuits and Signal Processing Series,

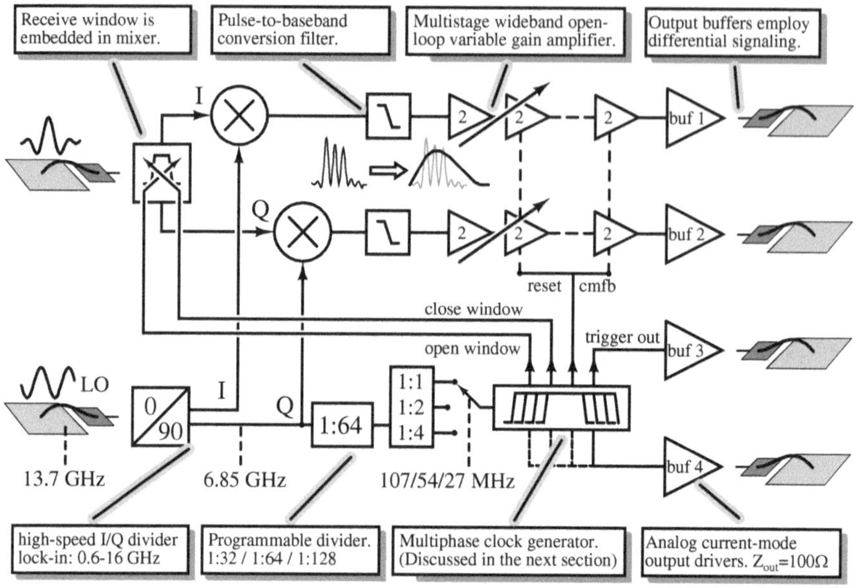

Figure 6.1. Schematic overview of the pulse-based receive unit. Note that in the actual design of the prototype receiver chip, the receive window was erroneously placed after the downconversion mixer, causing problems with in-band spurious noise.

of which is essential for the operation of the pulse-based system. Only during the receive slot assigned to a particular receive unit, the window is briefly activated in order to allow the receiver to capture some RF power from the channel. During the off-state of the receive window, unwanted noise and in-band interferer power is prevented from entering the signal chain, which significantly reduces the linearity requirements of the ensuing signal processing blocks. A time-domain filter approach is in fact the only viable option for a wideband receiver, because it guarantees a minimum of interferer suppression under all possible conditions, without the need for a (strictly controlled) bank of tunable analog notch filters. Based on the above described setup of the receiver, it is already possible to derive some specifications for the windowing circuit.

6.1 Receive window specifications

Because the windowing subcircuit itself is built around a non-ideal network of CMOS switches, it can only offer a limited performance in terms of insertion loss in the on-state and forward isolation during its off-state. Of course, one can make the ratio between the former two parameters very large, but it must not be forgotten that this structure is located within the very sensitive signal path of the receiver. The complexity of the windowing circuit should be kept to an absolute minimum in order to avoid capacitive coupling between the clock lines and the received RF-signal. The most critical factor here is charge injection due to the overlap capacitance between the steering gate and the drain/source terminals of the MOS transistors. The repetitive nature of the injected signal (caused by the opening and closing of the receive window) generates spurs in the signal path of the receiver and can cause clipping further on in the baseband amplifier.

The dimensions of the transistor switches should be kept as small as possible, with some obvious consequences for the isolation performance of the window. For example, a circuit can use a shunt switch to divert unwanted interferer power to a (virtual) ground node. The transistor that is used for this purpose has a certain on-resistance. This means that a certain amount of power leaks into the signal chain during the off-state of the receive window which results in implementation losses (IL).

However, a considerable part of interferer power leaks into the receiver anyway, during the on-state of the receive window. On average, the leakage power makes up only a small portion of the total interferer power at the output of the windowing circuit. If the duty cycle of the receive window is 10%, the suppression factor of narrowband interferers is 10 dB. It makes no sense to demand an isolation of the shunt transistors which is several orders of magnitude better than the suppression ratio of the time-domain filter itself.

For example, suppose that an implementation loss of 0.4 dB can be tolerated. It can be calculated that this boils down to the fact that leakage-induced power makes up about 10% of the total interference which enters during the active phase of a receive slot. It follows that the isolation of the switch must be $S_{21} = -20\,\mathrm{dB}$ (or better) with respect to the insertion loss in pass-through mode. Then, considering the impedances at the surrounding circuit nodes, the maximum value for the on-resistance of the shunt switch can be easily determined. Finally, based on the maximum on-resistance, the minimum dimensions of the shunt window switch can be determined.

IMPORTANT NOTICE!

A second measure that significantly reduces the effects of charge injection is to place the window switch in front of the downconversion mixer. In the design of the chip that is described here (see figure below), the window switches were incorrectly positioned at the outputs of the downconversion mixer. As will be shown later on in the measurement section, leakage of clock signals into the sensitive baseband section of the receiver resulted in spurious components at the symbol frequency and higher harmonics. Because of this, clipping occurred in the three final sections of the eight-stage variable gain amplifier. The issue was temporarily resolved by adjusting the internal routing table of the receiver in order to redirect the signal from an earlier gain stage directly to the output buffers. However, the sensitivity of the receiver was seriously degraded due to the problem of clock injection.

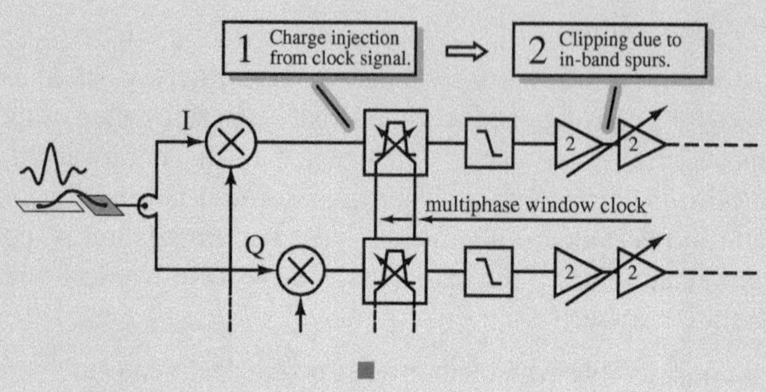

NOTE: The architecture with window switches behind the mixer is not further discussed in the remainder of this text.

By positioning the window circuit in front of the downconversion mixer, it is avoided that switching noise becomes an in-band interferer signal in the frequency band-of-interest. This is because the passband frequency of the received RF pulses is located between 3.1 and 10.6 GHz, while the main PSD lobe of the 1 ns-wide receive window is located in the frequency band below 500 MHz. As a result, charge injection from the clock signal directly into the signal band does not corrupt the signal-of-interest, the latter being a passband signal at the input of the downconversion mixer.

Of course, energy from the sidelobes of the receive window may also corrupt the received signal, so the clock signal which is driving the receive window should be low-pass filtered to smooth its rising- and fall slopes. Because the first notch of the receive window and the lowest RF signal band (3.1 GHz) are separated less than a decade in frequency, this is preferably done using a higher order filter with a steep rolloff. Removing some of the high-frequency information from the clock signal results in much smoother transitions from the on- to off-state of the window switches (Figure 6.2). It follows that the shape of the receive window is not rectangular any more, but has a rather sine-shaped profile.[1] The result is a reduced sensitivity near the edges of a receive slot.

[1] To be more precise, the shape of the clock signal is also multiplied by the transconductance characteristic of the CMOS switch itself.

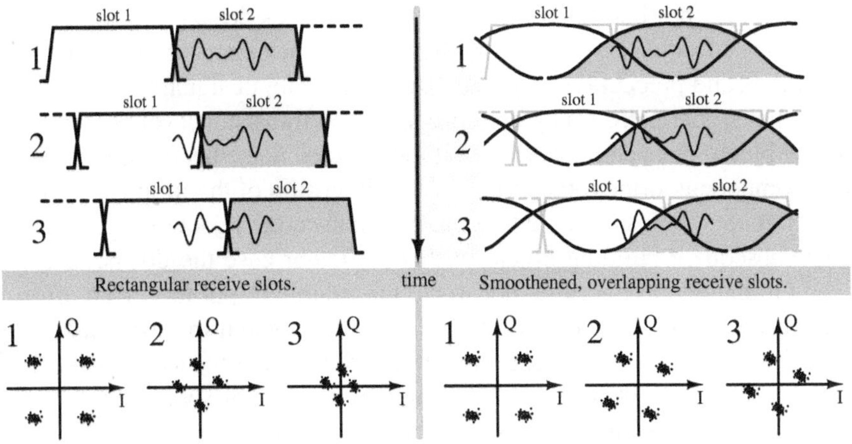

Figure 6.2. To cope with the transient effect of a multipath pulse stream shifting out of a receive slot as a result of clock offset, a smoothened receive window is used. This gives the phase tracking algorithm more time to compensate for the rotating constellation or to reallocate its resources to another slot.

This effect is in fact desirable, since it smooths out the transient effects of a pulse stream that slowly shifts out of the scope of a receive slot that is assigned to some particular receive unit. This can be understood bearing in mind that the baseband signal is formed by the combination of several pulses that fall within the scope of a single receive slot. The compound signal is formed by the vector sum of all these individual components, each with their own magnitude and phase delay. If for one reason or another (either due to changes in the length of the propagation path or due to clock offset), a strong multipath component shifts out of the receive slot, this will result in a change of both magnitude and rotation of the baseband I/Q constellation plane in that slot. In a receiver with a rectangular window shape the transition will be very abrupt, without giving the digital back-end sufficient time to react on this. For a window with smooth edges, the strong signal component will shift gradually out-of-scope, so the system has the opportunity to anticipate on this and correct for the varying rotation in the constellation of the symbol stream. To prevent the creation of *blind spots* in the search space of the slotted receiver, it is strongly advised to provide for some overlap between the positions of two neighbouring receive slots. This should also be taken into account by the back-end controller of the system, though, since the overlap invalidates the assumption of independent fading between receive slots. The system controller may avoid that overlapping slots become activated at the same time by different receive units.

Summarizing, the first countermeasure against clock leakage into the signal path is to use frequency planning and make the clock spurs an out-of-band issue. The second precaution to protect the weak RF input signal is to use a fully differential signal path. Using a well-considered transistor-level design of the window block, one can make sure that the injected noise becomes a common-mode component, orthogonal to the signal dimension of the RF pulses. The injected signal is then suppressed by the common-mode rejection ratio (CMRR) of the subsequent building block, in this particular case the downconversion mixer. On the other hand, it is not worth the trouble to put too much effort in the suppression of clock feedthrough. From the moment the overhead of the injected spurs has been brought down to approximately the same power level as this of the signal-of-interest, the remaining portion of switching noise can be removed by the signal processor in the back-end of the receiver. The repetitive nature of the receive slot causes that interferer energy stays concentrated around some discrete locations in the baseband spectrum. Mixing products between the clock signal and the local oscillator can be easily predicted while and affected frequency bands are subsequently repaired by the ISSR routine.

Remark that the problem of clock injection also works the other way around. In order to prevent that switching noise leaks back into the receive antenna, a direct path between the windowing circuit and the antenna terminals should be avoided. The same argument applies even more strongly to the LO-signal of the mixer, as this signal is within the frequency band of the antenna. Usually, in a narrowband receiver, reverse isolation is the responsibility of the low-noise amplifier (LNA). However, it may be more appropriate to use the word *antenna buffer*: for a wideband receiver with a considerable risk of high power in-band blockers, the linearity of the input buffer is at least as important as the noise figure of the input stage. Also, the noise figure requirements of a wideband system are more relaxed than this of a narrowband receiver, due to the larger share of thermal channel noise in the link budget.

6.2 Multiphase clock generator

The heart of the pulse-based receive unit is formed by a multiphase clock generator, used to orchestrate the interactions between the different subcircuits. All internal clocking is derived from an on-chip prescaler running at twice the LO frequency of the downconversion mixer. For practical reasons, the high-speed prescaler is injection-locked on the third-order harmonic of an externally supplied clock. The prescaler has a wideband pull-in range between 0.2 and 5.3 GHz, corresponding to an internal operating speed starting from 0.6 up to 16 GHz. The prescaler produces four 90° shifted LO signals (carrier frequency 0.3–8.0 GHz) which are applied to the inputs of the double-balanced quadrature mixer (see Section 6.3) and two divider blocks.

One of the dividers is part of a fixed 1:64 postscaler section (Figure 6.1), which is used to keep the baseband section of the receiver alive. The second divider is used as dummy ballast for the remaining output nodes of the high-speed prescaler. Failing to do this puts an unequal load on different outputs of the prescaler. This would result in I/Q imbalance errors between the 0°/180° and the 90°/270° inputs of the quadrature downconversion mixer. The output of the 1:64 postscaler[2] is further fed into a 3-bit programmable divider. The signal of this divider forms the heartbeat of the receiver and controls all time-critical functions, including timing of the receive window circuitry.

Before being further distributed to the different subcircuits, the heartbeat clock is converted to a multiphase clock signal. For this purpose, the output of the programmable divider is buffered by four parallel interconnected delay blocks. The delay blocks can be programmed independently from each other to delay the heartbeat clock between d_0 and $d_0 + 3\,\text{ns}$. The clock delay itself is accomplished in an analog manner, by varying the load between the outputs of a differential pair. The load itself formed by the $1/g_m$ impedance of a diode-connected MOS transistor, the current through which is controlled by the output voltage of a 4-bit on-chip DA-converter (Figure 6.3). In this way, the phase shift

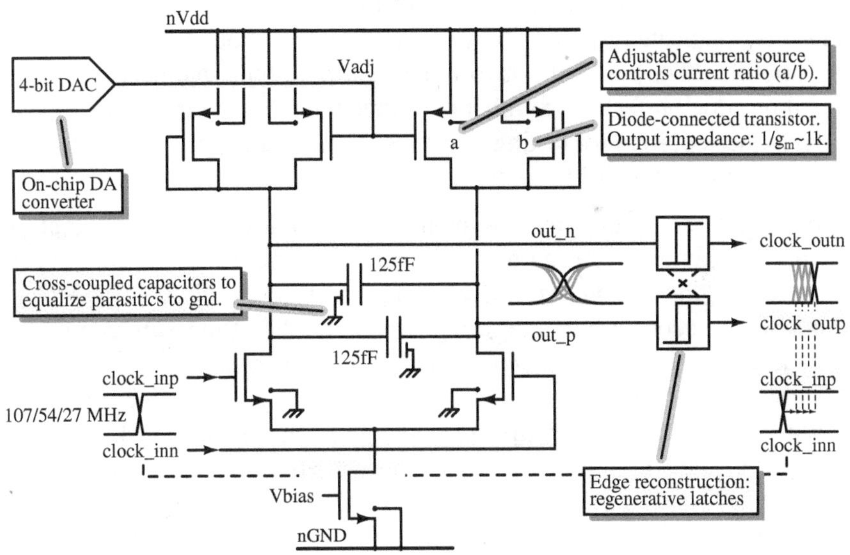

Figure 6.3. A single delay subcell of the multiphase clock generator. Four of these precision analog delay cells work in parallel in order to generate four independently controllable clock phases. The mutual delay between the first two phases marks the start- and stop-time of the receive window.

[2]The nominal operating frequency of the receiver is 6.85 GHz. The output of the 1:64 fixed postscaler yields the 107 Msymb/s sampling speed mentioned in the introduction of this section.

of the delay blocks can be programmed in 16 discrete steps, with a temporal resolution better than 200 ps. Because the varying output load of the differential pair affects the slope of the clock signal, the analog delay cell is followed by a regenerative latch. The result of all this tremendous effort is a four-way multiphase clock generator, with independently triggerable edges.

The first two clock phases are used to mark the opening and closing times of the receive slot: the rising edge of the first phase will open the receive window, while the rising edge of the second phase disables the window. This process repeats at the beginning of every master clock cycle. The third edge of the multiphase clock generator triggers the differential offset compensation circuit of the open-loop variable gain amplifier.[3] Finally, the fourth edge of the multiphase clock is used as a synchronous trigger signal for the (off-chip) AD-converters of the I/Q baseband section. This allows to dynamically adjust the optimum sampling point of the converter.

Measuring the aperture of the receive window

Connecting the pulse-based receiver to the outside world is done using four buffers. They are capable of directly driving the 50 Ω load impedance of external measurement equipment. Two of these unity-gain analog buffers are exclusively dedicated to the I/Q quadrature baseband outputs of the variable gain amplifier. The remaining two other analog output channels are used as general purpose buffers.[4] A number of interesting internal circuit nodes can be routed to these buffers for testing purposes. Their main application is to verify the correct functioning of the receiver.

Apart from some voltage measurements on internal nodes, the general purpose buffers can also be used to monitor the outputs of the multiphase clock generator. By connecting two different phases of the multiphase generator to the output, it becomes possible to perform some direct measurements of the phase delay between the respective clock lines. The accuracy of comparative measurements between two channels, however, is very susceptible to small differences in the connection to the external equipment. At this level, even small deviations of the length of the propagation path affects the practical accuracy. For example, bending the coaxial cables resulted in a relative time delay of about 10 ps. The same effects were also noticed at the RF inputs of the receiver, where slight bending of the cables resulted in a rotation of the baseband constellation of up to 20°.[5] Direct access to the clock lines of the multiphase

[3] Internally, the offset compensation circuit is based on the switch-cap topology, because it allows to realize a low-frequency pole without the need for an excessive amount of chip area.

[4] The general purpose buffers can be put to sleep to reduce switching noise on the supply voltage.

[5] Measured with $f_{LO} = 6.85\,GHz$. This only shows the importance of the phase-tracking mechanism of the signal processor, especially for the wireless channel: the differences in the propagation path length here are much larger and faster varying than in the artificial environment of a measurement setup.

generator is very useful to check the operational status of the chip, but the accuracy of direct measurements is plainly insufficient to yield reliable estimates of the aperture of the receive window.

Because of the aforementioned reason, an indirect approach was used to verify the effective aperture of the window. Remember that the windowing block is driven by an almost rectangular clock signal, which is low-pass filtered to suppress high-frequency components. The resulting receive window has a raised-cosine shape in the time-domain, of which the amplitude frequency response displays a decaying $\sin(x)/x$ shape. The damping ratio depends on the *rolloff factor* [Pro00] of the edges of the window. The output signal of the receive window is subsequently multiplied with the LO-frequency of the mixer, so the combined transfer function has a $\sin(x)/x$-shaped spectrum, centered at the LO-frequency of the receiver (see Figure 6.4). The null-to-null bandwidth of this frequency response can be directly related to the duration of the receive window. Although a direct measurement is not possible due to the bandwidth restrictions of the baseband section, the passband frequency response of the receive window can be determined by a set of narrowband measurements.

For this purpose, a narrowband carrier signal was injected into the RF input stage of the chip. Instead of an external signal that is being processed by the receiver, the sinusoidal carrier can be seen as the LO-signal which translates the frequency response of the pulse-based front-end itself to baseband. By

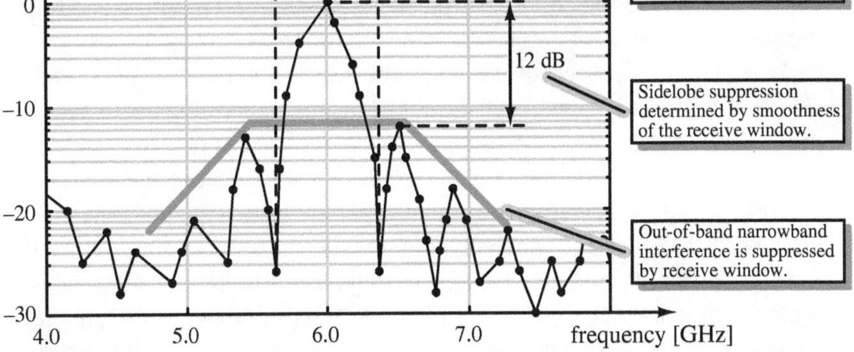

Figure 6.4. Indirect measurement of the receive window characteristics of the pulse-based received prototype. The null-to-null bandwidth of the main lobe is 650 MHz, corresponding to a window length of approximately 3 ns.

scanning the externally injected carrier over the frequency band-of-interest and measuring the integrated signal power in a small band around DC, a fairly accurate image of the ESD frequency response can be reconstructed. The cut-off frequency of the low-pass filter in the baseband section determines the frequency resolution of the scanning window. Figure 6.4 shows the test results of such a measurement, obtained with a resolution bandwidth of 12 MHz. Remark that the center frequency of the receiver is 6 GHz. The effective aperture of the receive window has been set to 3 ns, which can be deduced from the main-lobe bandwidth of 650 MHz. The absolute level of the energy spectral density (ESD) in Figure 6.4 depends on several parameters, such as the externally injected carrier power, the RMS aperture width of the receive window, the gain settings of the variable gain amplifier and the resolution bandwidth used during the measurement. Because the absolute peak level is of no further interest for this discussion, the frequency response has been normalized to the peak ESD at the center frequency of the receiver.

– SENSITIVITY DETERMINED BY THE RECEIVE WINDOW –

It is important to recognize that the bandwidth of the main lobe alone is not a correct measure for the total amount of RF energy that is captured by the front-end or the sensitivity of the front-end in a particular frequency band. In contrast to what would be expected, the sensitivity of the receiver improves with a reduced null-to-null bandwidth of the main lobe.

Intuitively, this can be understood by realizing that if the duration of the receive window is increased – which is in fact the only way to reduce the bandwidth of the main lobe – the receiver indeed allows more energy to pass to the signal chain. Mathematically, this finding can be verified by understanding that the peak magnitude of the ESD response is proportional to the square of the window length, which makes that the total area (energy) under the main lobe increases more or less linearly with an increased duration of the receive window. The reader should also be fully aware of the fact that the spectral shaping technique used to determine the frequency response of the receive window does not apply to wideband input signals. This is because the spectral footprint of the received signal must be *convoluted* ($\otimes$) with the frequency response of the receiver. It is an easy mistake to suppose that the spectrum of the received signal is shaped (multiplied) by the magnitude frequency response of the receiver. Keeping this in mind, it can be seen that there is nothing suspicious in the magnitude frequency response of the pulse-based receiver being smaller than the spectral footprint of the received pulse stream.

Also interesting to note is that the sidelobe levels outside the passband of the main lobe are reduced to below -10 dB. This implies that the susceptibility for narrowband out-of-band interference is improved by an additional factor of 10 dB, on top of the suppression ratio already offered by the duty cycle of the receive window. In theory, dynamically controlling the shape of the receive window offers some perspectives to position one of the nulls of the receiver response at a frequency point with high interferer power. However, the practical implementation of this method can be very complicated, for only a marginal revenue and a marginal flexibility. To conclude, observe that the spectrum plot of Figure 6.4 shows a decreasing trend line. This is due to the limited frequency response of the LQFP-32 package that was used for the test setup.

6.3 RF input stage

The architecture of the prototype chip has already been introduced in Figure 6.1. The circuit implementation includes a two-way I/Q signal path, of which each part is constructed of a window block, a pulse-to-baseband downconversion mixer, a baseband variable gain amplifier and a set of analog output drivers. The most performance-critical sections are located in the RF input of the receiver. Figure 6.5 shows the basic structure of the downconversion mixer used in the prototype version (the supporting biasing circuits are omitted). The mixer is based on a double-balanced CMOS Gilbert cell [Gil68] with modified input stage. Remark that the differential input signal is injected in parallel with the current sources of the mixer. The cascode transistors between the injection point and the mixer switches have a twofold purpose.

Their first task is to provide a low input impedance, as seen from the entry point of the RF antenna signal. The result is a wide input bandwidth, while most of the incident RF current is effectively forced to flow through the switches of the mixer. The second goal of the cascoded input stage is to isolate the mixer and the receive window circuitry from the input of the receiver. This is to prevent that LO-spurs and clock harmonics become unintentionally radiated by the receive antenna. The differential output of the mixer is shorted by a capacitor, which is part of a low-pass filter and plays an important role in the pulse-to-baseband conversion process (see Section 5.1). The resistive part of this low-pass filter is formed by the real part of the mixer output impedance. Part of this output impedance can be accounted to the PMOS current sources on the top of the circuit, while another part is caused by the effective resistance of the four cross-coupled switches during their crossover period.[6]

[6]The actual value of the crossover impedance depends on the amplitude, frequency and slope of the LO-signal applied to the mixer switches and should be determined using transient simulations.

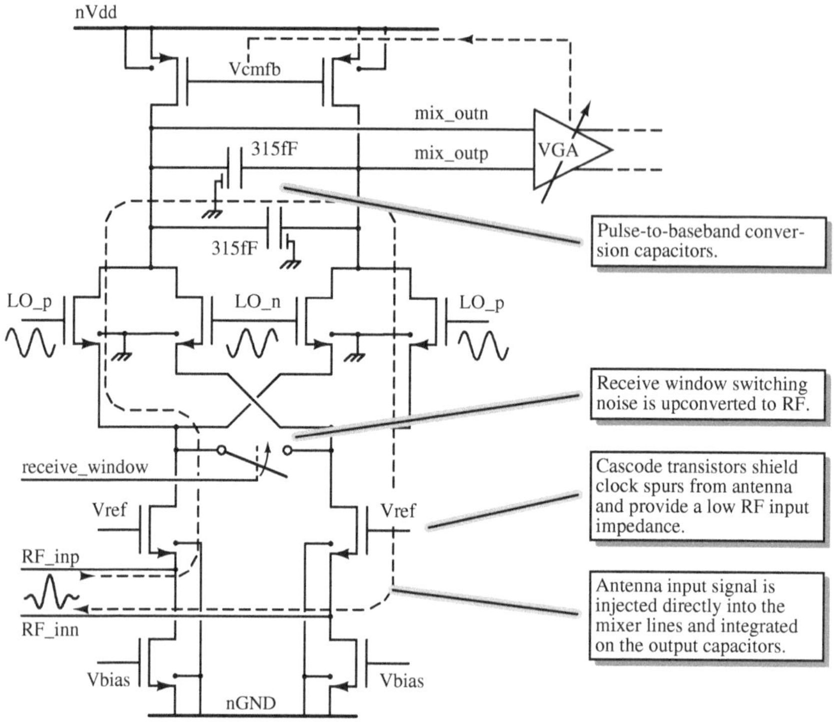

Figure 6.5. The input stage and downconversion mixer of the prototype chip are based on a generic Gilbert mixing stage. Remark that the RF input signal is injected directly into the mixer lines, for improved linearity performance (IIP3 better than −1 dBm).

For linearity reasons, the prototype receiver does not feature an active low-noise amplifier at its input, so the RF input signals are directly delivered to the mixing stage. The result is that there is no active gain involved in the pulse-to-baseband conversion process. It is important to realize that, during the pass-through state of the receive window, the injected RF current is routed through the mixer switches and is then integrated on the capacitor between the output terminals. The only parameter that matters at this point is the charge that the signal-of-interest induces on this capacitor. The thermal noise floor measured across this capacitor is determined by $v_{\text{noise, RMS}}=\sqrt{kT/C}$ [Sar93], but its capacitance is not relevant for the SNR of the output signal. This is because the capacitor samples a current that is being injected in the RF input, rather than a voltage value from a low-impedance voltage source. The pulse energy stored on the capacitor is independent from its value, because the total charge that is extracted from a single time-limited RF pulse applied to the input is limited to $Q_{tot} \approx i_{injected} \cdot T_{pulse}$. While it is true that increasing the capacitance would reduce the thermal noise voltage,[7] the voltage related to the signal-of-interest will decrease by the same factor.

The reader may correctly argue that the above reasoning is only valid as a first order approximation. For example, it was supposed that all of the RF signal current is integrated on the capacitor. It is clear that this may not entirely reflect reality for frequencies below the passband of the low-pass filter, when the output impedance is not longer dominated by the capacitor, but by the output resistance of the mixer. For frequencies below the cut-off frequency of the filter,[8] some of the signal energy leaks away into the resistive part of the filter. However, only a small portion of the signal is affected by this problem, since most of the spectral density of the pulses is spread over a much larger bandwidth than the passband frequency of the low-pass filter.

For example, if the bandwidth of the pulses is 1 GHz, while the cut-off frequency of the filter is 50 MHz (for a symbol rate of 100 Msymb/s), the signal loss is limited to about 0.2 dB. This is the price that must be paid for not using an ideal matched filter based pulse-to-baseband converter. It might be tempting to improve the efficiency of the input stage by moving the pole of the filter towards a lower frequency and later on, in the signal processor, compensate for the distortion (caused by the non-flat frequency response) of the baseband signal. Doing so does not necessarily improve the performance: a filter with a low frequency pole can be considered as an integrator. In an open-loop configuration, low-frequency components will cause large deviations from the average signal level. The large dynamic range between low- and high-frequency components of the baseband signal has an obvious impact on the loading factor of the AD-converter. The increased dynamic range is not a problem for the signal-of-interest, but in-band interferers could clip if the signal headroom of the AD-converter is not scaled accordingly.

Measuring the noise figure of the front-end

The output of the mixer is connected to a variable gain amplifier. The VGA itself is based on a chain of eight 7.8 dB-gain segments. The gain of the VGA can be controlled by rerouting the output of one of these segments directly to the output buffer of the chip. Power consumption is further minimized by switching off unused sections of the amplification chain. The individual amplifier sections are based on an open-loop amplifier topology, with some special precautions to suppress third-order harmonic distortion (HD_3). This topic is covered in greater detail in Appendix 7, so the specific implementation aspects of the

[7]Only circuit-induced thermal noise is considered here, not the background noise of the channel.

[8]After downconversion of the RF pulses by the mixer, low frequencies emerge in the (wideband) spectrum of the output current.

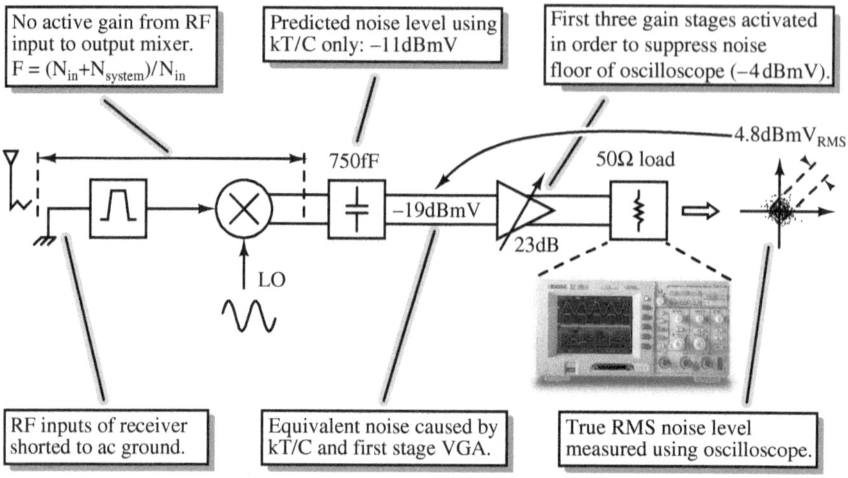

Figure 6.6. Sensitivity analysis of the prototype receiver. In order to measure the noise produced by the receiver itself, the RMS noise level at the output was measured while the RF input terminals were shorted to AC-ground. The actual noise figure of the system depends on the background noise of the channel.

baseband amplifier will be ignored for now. For more details about the open-loop baseband amplifier, the reader is referred to Section 6.7.

The goal of this subsection is to determine the noise produced by the analog front-end of the receiver. It is expected that most of the noise power originates from kT/C noise at the output of the mixer, plus a limited amount of thermal noise being produced by the first few stages of the variable gain amplifier. In order to exclude noise originating from the channel, the RF inputs of the receiver were first shorted to AC-ground. Then, the RMS noise signal magnitude at the output of the baseband section was determined using a digitizing oscilloscope (Figure 6.6). The total measured noise power was −42.2 dBm (4.8 dBmV measured over a 50Ω load), while the integrated noise floor of the oscilloscope itself was found to be −56 dBm (−4.6 dBmV). At the moment of the measurement, three stages of the variable gain amplifier were activated, which corresponds to a total voltage gain of 23.4 dB. As a rough estimate, it can be assumed that the equivalent noise amplitude over the terminals of the capacitor at the output of the pulse-to-baseband conversion mixer should be about −19 dBmV. Using the theoretical kT/C noise formula, the RMS noise voltage across this capacitor (value 750 fF) can be calculated as (6.1):

$$v_{\text{noise, RMS}} = \sqrt{\frac{kT}{C}} = -23\text{dBmV}\,, \tag{6.1}$$

$$\text{with } k = 1.38 \cdot 10^{-23}\text{J/K},\ T = 300\text{K and } C = 750\text{fF}$$

It can be concluded that the experimental results fit the theoretical noise level fairly well. Note that the actual measured noise level is slightly higher than the theoretical noise value. This deviation can be accounted to the thermal channel noise of the input pair in the first stage of the VGA, which was ignored in the above analysis. Note that the capacitance used in (6.1) does not only include the physical capacitor at the output of the mixer (630 fF), but also the output impedance of the mixer and the load of the VGA ($\approx$120 fF). One way to reduce the equivalent input noise of the baseband amplifier is to increase the transimpedance of the first input stage by increasing the width of the input pair. The higher load from the input pair must be compensated by a smaller capacitor at the output of the mixer. Remark that it is not a good idea to replace the entire physical capacitor by the capacitive load of the VGA. The underlying problem is that the parasitic capacitance between the differential inputs of a differential CMOS transistor pair has a parasitic capacitance to AC-ground, due to the drain capacitance of the current transistor at the common source node. For high frequency components in output current of the mixer (above the cut-off frequency of the baseband amplifier), this would imply that the mixer is loaded by two grounded capacitors, which would destroy the benefits of a fully differential architecture.

The only thing left over to specify is the *noise figure* of the analog front-end. For this, let us first define the *noise factor* of a component in a system. The noise factor (F) is defined as the ratio between the SNR of the signal at the input versus the remaining SNR value at its output. Since a circuit block can only add noise to a signal, the noise factor is always larger than unity. When the noise factor is expressed in decibels, it is referred to as the noise figure (NF) of the system (6.2):

$$\begin{aligned} \mathrm{F} &\triangleq \frac{\mathrm{SNR_{in}}}{\mathrm{SNR_{out}}} > 1 \\ \mathrm{NF} &\triangleq 10\log_{10}(\mathrm{F}) > 0\,\mathrm{dB} \end{aligned} \quad (6.2)$$

Earlier in this section, it was pointed out that the input stage does not offer any active gain, since the RF input signal is directly injected into the mixer input lines. This was done to achieve wideband input matching and to obtain a high linearity of the input stage. For a system block without active gain, the formula of the noise factor merely reduces to a ratio between noise power levels (6.3):

$$\mathrm{F}_{[\text{no gain}]} = \frac{\mathrm{S_{in}}/\mathrm{N_{in}}}{\mathrm{S_{out}}/\mathrm{N_{out}}} = \frac{\mathrm{N_{out}}}{\mathrm{N_{in}}} = \frac{\mathrm{N_{in}} + \mathrm{N_{system}}}{\mathrm{N_{in}}} \quad (6.3)$$

It follows that in absence of interference, the thermal background noise captured by the antenna is the only factor that contributes directly to the input noise of the system. For example, when the RF input bandwidth of the receiver is 1 GHz, the noise floor at the antenna terminal is given by kTB = −83.8 dBm. If the duty cycle of the receive window is 10%, the total channel noise power that is allowed to enter in the front-end of the receiver is −93.8 dBm. At the output of the mixer, the kT/C noise power is approximately −85.6 dBm.[9] From this, it can be calculated that the noise figure of the receiver is NF = 8.2 dB. This is a rather high value which is caused by the absence of an LNA in the prototype receiver. This noise figure can be improved by putting a low noise amplifier in front of the receiver. For example, an LNA with a noise figure of 2 dB and a gain of 3 dB would reduce the overall noise figure of the receiver to 4.4 dB.[10] However, adding an LNA to the front-end should be done with great caution, because the LNA is located in front of the receive window and is exposed to the full interferer power. In a hostile channel with high interference levels, an LNA may cause more harm than good [Ver06]. The prototype receiver itself, without active gain element at the input, exhibits an input-referred third-order interception point (IIP_3) better than −1 dBm.

6.4 Design for testability

During the design of the receiver, considerable time and effort was put in the testability of the system. The prototype receiver has a reasonable number of internal nodes and voltage settings for which it would be interesting to have external control lines or an access point for measuring purposes. For this reason and to prevent excessive complexity in the measurement setup, the receiver has been equipped with an on-chip memory bus and an array of analog and digital supporting circuits. With this approach, over 50 in- and output nodes could be successfully merged in a resource-efficient measurement shell around the core of the receiver.

The framework of the control system is based on an on-chip static memory, consisting of a number of stacked memory cells. Each of these single-bit cells is implemented by a true single-phase (TSP) D-flipflop. Such a TSP-based cell has – apart from power supply – only three connections: one input data bit, an output bit and a single clock line, which is shared among all memory cells. Data bits are shifted in a serial way from one memory cell into the next, so the interconnect pattern between the cells is limited to only two wires. Instead of being centrally located, the memory bus is divided over a number of small 4-bit subblocks, spatially distributed over the floorplan of the chip. This way, the data outputs of the memory bus could be located precisely where needed, preventing that wide bundles of wires 'fan out' from the memory unit to specific locations elsewhere in the receiver.

[9] The real part of the mixer output impedance is 2 kΩ. Using Equation (6.1), calculating the RMS noise power is straightforward.

[10] Noise factor of a cascaded system: $F_{tot} = F_1 + \frac{F_2-1}{G_1} + \frac{F_3-1}{G_1 G_2} + \cdots$

At the next higher level in hierarchy, the output of one memory subblock connects to the data input of the subsequent block, which results in a chip-wide serial register. The input of the first register is made available as an external pin, as is the clock line of the memory bus. The system bus is controlled by an external microcontroller, which applies the correct pattern of logic signal levels to the data and clock lines in order to shift a vector of settings into the receiver. The output of the memory bus is also made available on an external 'data out' pin,[11] which allows the controller to read back information at the output of the memory bus. A control word is always read twice into the memory bus. In this way, the memory controller checks for transmission faults and flags an error to inform the user about this problem. For all this, the controller only needs to know the length of the settings vector and the sequence of activation of the control signals to clock a data bit into the memory bus. However, it does not know about the structure of the memory map of the receiver.

For this purpose, the controller communicates with a computer over an optically isolated RS-232 interface (Figure 6.7). On the host computer, the user is presented a user interface displaying the chip floorplan of the receiver (Figure 6.8). On this floorplan, the structural blocks (mixer, VGA, multiphase generator) of the prototype receiver are visualized. Along with each block, the user is also presented with a number of interactive controls related to that specific block. For example, the mixer block contain a GUI slider which allows to control the bias voltage of the cascode transistors. Both the I- and the Q-mixer are fed by the same antenna signal. Adjusting the mutual input impedance between the mixers by varying the gate voltages of their respective cascode transistors alters the ratio of how incoming signal power is distributed among the two stages. This is an interesting possibility in light of the pulse-based wideband receiver discussed before, where the power from a single physical antenna has to be split among multiple receive units.[12] Another example is the variable gain amplifier. The user can select the number of VGA-stages that are activated, while the total gain of the current setting is displayed for operational convenience. In the background, the gain control is translated to the correct settings of the internal multiplexers of the receiver. The settings are subsequently packed in a new control vector, handed over to the microcontroller which transfers it to the memory bus of the prototype chip.

A similar strategy is used to adjust the delay lines of the multiphase clock generator. For the configuration of this block, the memory bus drives four on-chip

[11] The data output pin of the memory bus has an internal weak pull-up resistor. This pin is multiplexed with an emergency master clock input for the multiphase clock generator.

[12] Remark that this is in sharp contrast with narrowband architectures. To a certain level, power is less of a problem here since the surplus input impedance (gate capacitance) formed by multiple parallel stages can be taken into account in the matching network of the antenna. Tuning out load capacitances is not an option for wideband networks, though.

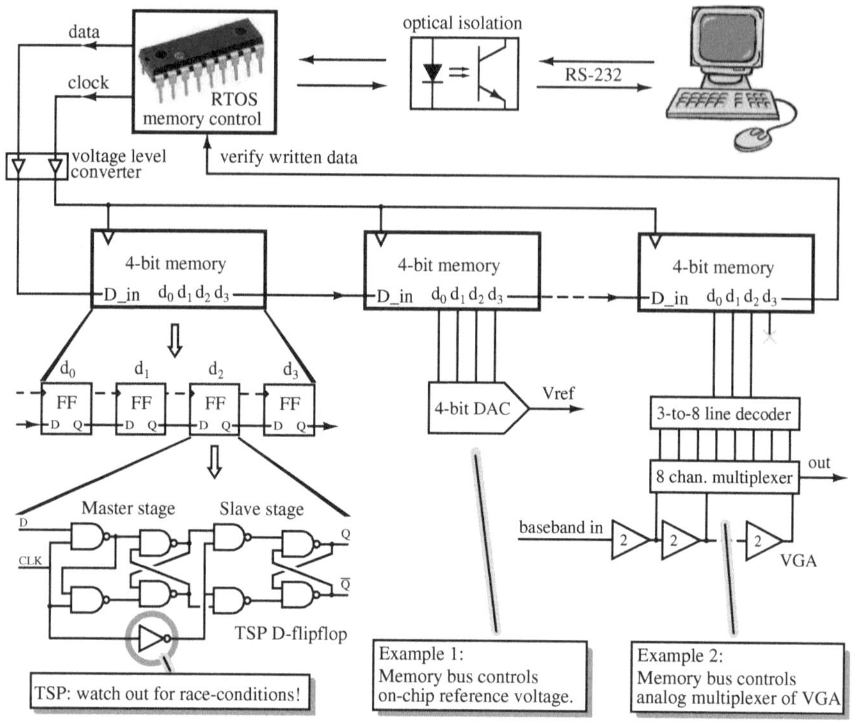

Figure 6.7. The prototype receiver is controlled by an on-chip memory bus which allows to control and measure over 50 in- and output nodes. The memory bus itself is controlled by an external microcontroller. The controller, in his turn, is connected to a computer which presents a convenient interface to the user.

binary coded DA-converters. The output signal of the binary DAC's is then used to alter the load impedance of the analog delay cells (see Section 6.2). It is clear that the user interface prevents that the user is distracted with the technical details of the measurement setup, while it becomes very convenient to load a set of predefined test configurations on a number of prototype setups. Also, settings can be linked to interesting measurement results and stored for further analysis later on.

Thanks to this programmable measurement shell which was built around the prototype receiver, the number of access points was reduced from 88 to only 22 bonding pads, without having to compromise on the accessibility of internal circuit nodes. The only external components required to do this are the three-wire memory bus, an external reference voltage from which the common mode voltage of the signal chain is generated and a single bias current for the embedded AD-converter circuits. The connection to the outside world is managed

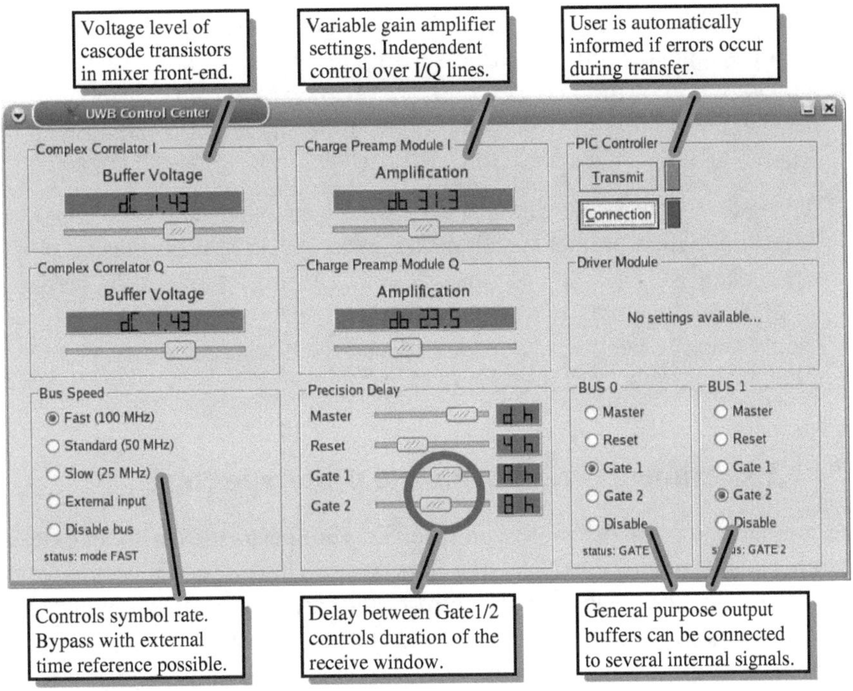

Figure 6.8. The GUI interface of the prototype allows the user to quickly load settings into the chip, compare the performance of different chip samples and store settings for future reference.

by four output buffers, two of which are dedicated to the analog signal chain while the remaining two serve as general purpose drivers for measuring internal voltage levels and (baseband) analog and digital signals. To reduce transients on the supply lines, a current-mode differential topology was chosen for the output buffers, even for digital signaling. Finally, off-chip RF baluns (0.4–450 MHz, 2:1 ratio) are used for the differential to single-ended conversion, to provide galvanic isolation and to transform the 50 Ω impedance of the load to the output impedance of the on-chip drivers.

BAD SCIENCE AT IT'S BEST

During the first debugging tests of the prototype receiver, a minor but practical problem was experienced with the on-chip memory bus. It seemed as if the contents of the static memory got corrupted about once or twice in an hour. More regular updates of the memory contents could resolve this problem most of the time, but intermittent failures continued to occur. After some digging, it turned out that the culprit was the high input impedance of the clock line between the external controller and the on-chip memory. A few centimeters of interconnect wiring is able to build up a sufficient amount of voltage noise to trigger the clock line of the memory once in a while. As a result of this, the data vector is shifted by one or more positions, thereby invalidating all internal settings.

A possible solution to this are differential encoded data- and clock lines. In the measurement setup of the prototype, the mysterious memory corruption issue (MMCI) was solved by placing a termination resistor close to the clock input lead of the receiver. An external interferer cannot provide the power needed to reach the threshold voltage of the clock line. A time consuming mistake!

6.5 Experimental results for the prototype chip

The measurements of the prototype receiver were performed with an external reference clock running at 2.25 GHz. The on-chip high-speed prescaler of the receiver is injection-locked on the third-order harmonic of this signal, which results in an internal reference clock of 6.75 GHz. The off-chip oscillator signal is presented to the prescaler in a differential format. This allows the user to fine tune the accuracy of the quadrature outputs of the prescaler. At the output of the prescaler, the four 90° shifted signals are applied to the LO-inputs for the I/Q downconversion mixer network. This makes that the center frequency of the pulse-based receiver is exactly 3.375 GHz.

Since the prototype chip includes a receiver, but lacks the hardware at transmission side for a fully operational wireless link, the signal applied to the RF inputs was generated by a multi-port data generator.[13] Two of the medium-speed ports of this generator were used to generate the differential clock signal (2.25 GHz), while a third high-speed port was used to generate a 13.5 Gbps serial bit stream. The accuracy of the latter one is sufficient to generate a test signal with a main lobe frequency at 3.375 GHz and four different possible phases with respect to the LO reference signal. The internal data registers of the generator were programmed to rotate the phase of the test signal between two consecutive receive slots, which results in a QPSK modulated carrier signal. A pulse-like QPSK transmission was achieved by limiting the duration of a pulse burst to 20 periods (40 data bits, period $T_{pulse} = 3\,ns$).

Remark that the signal of the data generator is not entirely compliant with the requirements prescribed by the FCC (see Section 6.5), which is for several reasons. First of all, part of the sidelobe power oversteps the spectral mask imposed by the FCC due to the steep edges of the rectangular transmit window.

[13] Agilent ParBERT 81250 with 13.5 Gbps modules.

Then, the RF signal also has a repetitive nature, caused by the generator looping over a limited block length of the internal data register. As a result of this repetition, the power of this RF signal is not evenly spread over the frequency band, but is clustered around discrete frequencies. The spectral lines are separated by a distance $1/T_{repetition}$, but also the peak PSD is scaled with the distance between two consecutive lines ($T_{repetition}$), possibly violating the spectral mask if the total transmission power is kept constant. In a regular pulse-based transmission, this problem is avoided by introducing sufficient randomness in the baseband QPSK symbol stream. In most cases, this condition is easily met when the encoded bit stream is scrambled with a pseudo-random noise (PRN) vector.

However, the issue above is not relevant for the outcome of further measurement results. The I/Q quadrature outputs of the receiver are connected to a digitizing oscilloscope in X–Y mode. The external trigger input of the oscilloscope is connected to one of the general purpose drivers of the receiver, and internally linked to the trigger channel of the multiphase clock generator (see Section 6.2). Triggering on the baseband signal itself is avoided because this results in unstable triggering points when no signal is present in the baseband output of the receiver. The synchronous trigger signal of the receiver ensures that the oscilloscope always starts to acquire data at a fixed phase angle, even if no signal is present. This situation typically occurs at the startup of the measurement setup. At that moment, chances are quite slim that the receiver is able to immediately pick up a signal from the data generator. The reason for this is that the pulses from the data generator are not necessarily synchronized with the single receive slot of the prototype pulse-based receiver. Unfortunately, the prototype chip also lacks any form of synchronization mechanism, as this is the responsibility of the digital back-end processor (see Section 5.2). Therefore, after the startup of the receiver, the RF signal from the data generator needs to be aligned manually until the transmitted pulses come within sight of the activated receive slot. Because a reallocation of the receive slot itself is not currently supported by the prototype receiver, this is done by delaying the RF stream from the data generator with respect to the oscillator clock. The multiphase clock generator (which controls the position of the receive window) is synchronized with the on-chip high-speed prescaler, while the prescaler itself is (injection-) locked on the clock of the data generator. The RF pulses will thus stay within scope of the receiver for as long as the receiver remains locked on the data generator.

The result of all this effort is shown in Figure 6.9. It shows the I/Q constellation plot of the received RF pulse stream after pulse-to-baseband conversion to a regular QPSK signal by the pulse-based radio receiver. Remark that the constellation points are not located exactly on top of each other, but form small *clouds* around the ideal constellation points. The deviation from the ideal

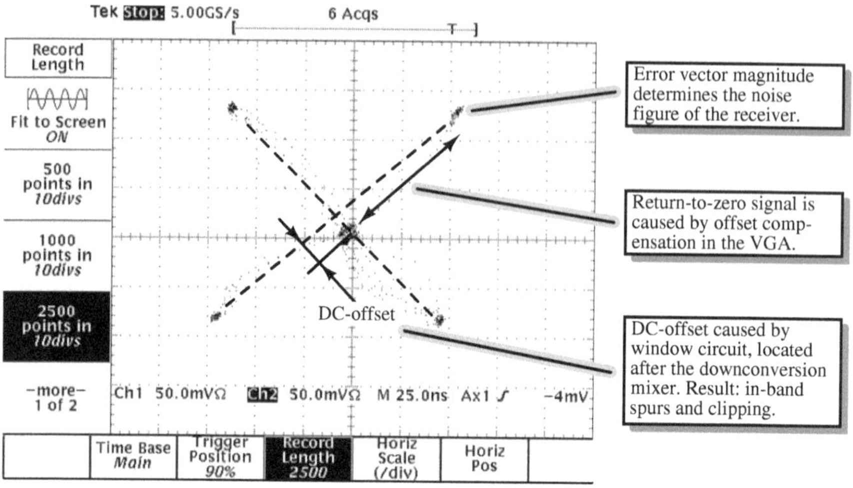

Figure 6.9. QPSK constellation result captured from a live measurement of the pulse-based prototype receiver. Remark the DC voltage offset of the constellation center. For weak signals and high VGA gains, this causes clipping further on in the receive chain which in turn reduces the sensitivity of the receiver.

reference constellation point is defined as the error vector. The error vector magnitude (EVM) is obtained by scaling the RMS power of the error vector to the RMS power of the received QPSK signal. In the absence of channel noise, the EVM can be used to measure the performance of the receiver front-end. This topic is discussed in more detail by [Has97].

During operation over a live wireless link, there is a direct relationship between thermal channel noise, the EVM and the raw BER performance of the receiver. The situation becomes a little bit more complicated for narrowband interference: if the baseband QPSK symbol stream is first patched up in an ISSR decoder before the demapping stage, the BER will depend on two other factors: the desensitization of the receiver as a result of the back-off from the ideal operating point and the equivalent amount of noise power added to the signal due to the removal of the affected frequency bands. The final figure here depends on a lot of factors (such as the gain setting of the baseband section, the bit depth and loading factor of the ADC), which makes that numerical simulations are more appropriate than complex theoretical calculations. The author decided that such high-level system simulations are out of the scope of this work, though.

Also noteworthy to mention is the return-to-zero phase between two consecutive samples of the receiver (Figure 6.9). This behaviour is caused by the regular interventions of the differential offset compensation circuit in the signal chain. Because the open-loop baseband amplifier in the prototype receiver uses DC-coupled sections, a growing offset voltage builds up after a few stages. In the prototype chip, this issue was solved by sampling the differential offset voltage on a large interstage coupling capacitor after every two sections of the baseband amplifier. However, the bandwidth requirements of the amplifier are doubled in this RZ approach.[14]

In the implementation of the improved baseband amplifier of Section 7, this problem was circumvented by the introduction of two independent common-mode feedback (CMFB) circuits on each differential output node. In this approach, the return-to-zero format is avoided, but signal frequencies within the control range of the CMFB circuit will be suppressed. Fortunately, ISSR can be used to reconstruct the missing portion of the signal. For example, when the offset compensation circuit cuts in below 1 MHz, while the bandwidth of the baseband signal is 50 MHz, the implementation losses are less than 0.09 dB. Also remark that in Figure 6.9, the QPSK constellation is not symmetrically centered around the origin of the I/Q plot. This is because the window circuit is erroneously located behind the downconversion mixer, which causes charge injection within the frequency band of the baseband amplifier.

Not visible in Figure 6.9, but clearly present in the spectrum plot of the baseband signal is a large spectral component at the clocking frequency of the offset compensation circuit. Part of the DC offset voltage in the baseband signal chain is in fact caused by self-mixing of the offset compensation clock signal. The consequence is that it cannot be removed by simply ac-coupling the stages of the amplifier. The important lesson learned here is to stay out of the frequency band of the signal-of-interest, especially when dealing with very weak signal components as those encountered in the signal chain in the front-end of a receiver (see Section 6.1).

6.6 Summary of the pulse-based receive unit

To conclude this section, Figure 6.10 shows the chip microphotograph with an overlay indicating the different building blocks of the receiver. The core area of the receiver measures 930 × 930 μm. Clearly visible on the floorplan of the chip is the quadrature signal chain of the prototype receiver. The differential RF signal inputs are located at the left side, while the baseband outputs (I/Q) can be found at the right side of the chip. From the left to the right, one can distinguish the downconversion mixers with integrated windowing circuitry, a dual baseband variable gain amplifier and the output driver section. The bottom section

[14]The minimum sample rate of the ADC is not affected if the converter is synchronized to the sample speed of the receiver using the multiphase clock trigger signal. However, for an asynchronous conversion technique, the bandwidth of the ADC would be more than doubled.

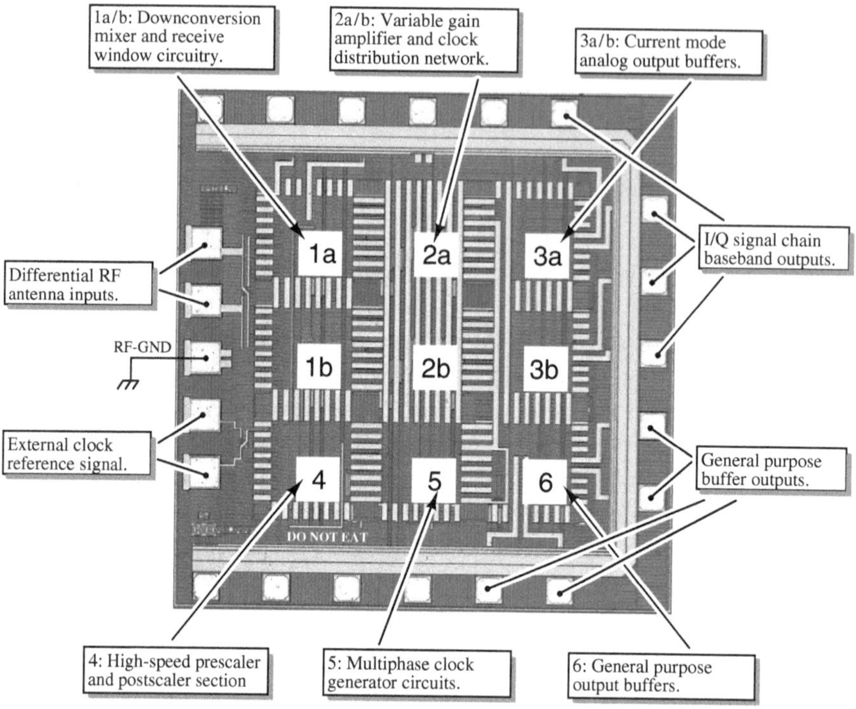

Figure 6.10. Chip microphotograph of the pulse-based prototype receiver. Each of the nine subblocks is surrounded by a power supply and decoupling ring, underneath which a ring of substrate contacts prevents that clocking noise leaks into the analog subsection.

of the chip contains at the left subblock (4) a high-speed prescaler (operation of which was verified up to 16 GHz) and the divider chain. The middle block (5) contains the multiphase clock generator, which is used as accurate time reference for the receive window circuitry. The outputs of the delay cells of the clock generator are buffered by regenerative driver latches. The latter ones are necessary to drive the capacitive load of the clock distribution lines, which are clearly visible as the 4×2 vertical lines protruding upwards from the multiphase subblock (5) into the analog signal chain (2a/b).

Remark that the clock lines are routed perpendicular to the I/Q signal chain in order to minimize in-coupling clock noise. For the same reason, the clock lines organized in pairs of differential signal conductors. Also, the distribution lines have equal lengths to equalize their load on the clock drivers. The microphotograph also shows the power distribution grid which surrounds each of the nine circuit subblocks. Embedded inside this power grid are the supply decoupling capacitors, for a total distributed on-chip capacitance of 45 fF. Wide diffusion guard rings were placed in the substrate directly underneath this low-impedance AC-ground plane in order to isolate the subblocks from each other. It is the responsibility of the power grid to provide a low-ohmic ground and supply reference plane to all subcircuits of the receiver.

The prototype of the pulse-based radio unit was implemented in a 0.18 μm standard CMOS process and has a supply voltage of 1.8V.[15] The total current drawn from the external supply source is 67 mA (120 mW). The largest share of the power consumption is taken into account by the high-speed prescaler and CML divider section (46%), followed by the four 50 Ω output drivers (29%) and the baseband amplifier (20%). The high-speed prescaler was conservatively overdesigned and a more conservative design should be able to bring down the power consumption of this block about an order of magnitude. The exact figure depends on the number of receive units that are driven by the prescaler: the reader should keep in mind that the prototype chip only accommodates one single receive unit. However, the pulse-based radio system – introduced in Section 6.5 – relies on a small set of parallel receive units, which increases the capacitive load on the quadrature LO-outputs of the prescaler. Furthermore, a significant amount of power could be saved in a fully integrated receiver design. The baseband instrumentation driver section can be omitted entirely if the variable gain amplifier is directly connected to an on-chip AD-converter.

The first test-chip (Figure 6.11) presented here has proven the feasibility of pulse-based radio receivers and at the same time brought a very important aspect of pulse-based radio to light: the very weak signals which are present in the first input stages of a receiver are extremely vulnerable to noise and clock injection. A direct connection (even capacitive) between noise sources containing spectral components within the frequency band of the signal-of-interest must be avoided under all circumstances. In the proposed receiver architecture, this goal is achieved by relocating the windowing switches to the input of the downconversion mixer. Frequency components within the RF signal band of the receiver (3.1–10.6 GHz) must be removed from the signal driving the windowing circuitry. Receivers based on template correlation will be victim to a considerable amount of in-band noise if the correlation is done early in the receive chain or will consume a lot of power due to the wideband signal amplification chain in front of the correlator section. Such implementations should always be regarded as highly suspicious, either in terms of sensitivity or in terms of power consumption.

[15]The nominal threshold voltages for nMOST and pMOST were about $V_{th,n} = 0.52$ V and $V_{th,p} = 0.48$ V.

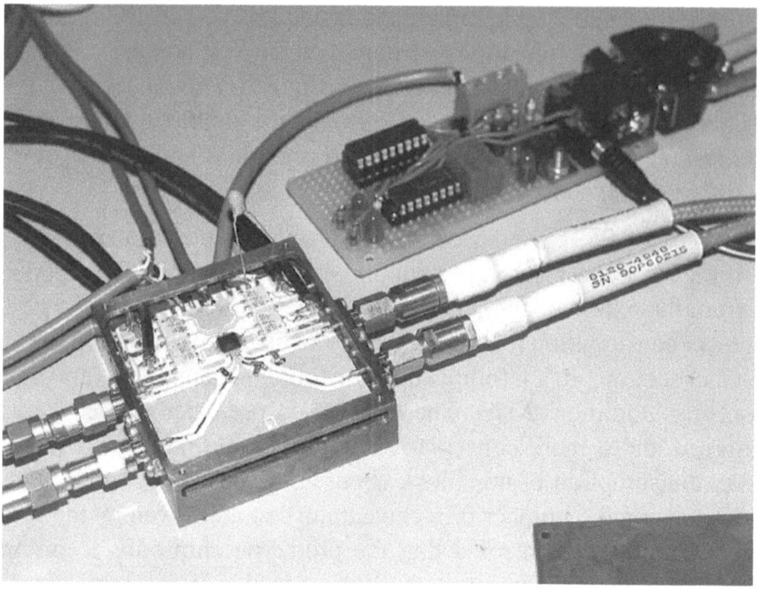

Figure 6.11. Photograph of the measurement setup. In the background is the microcontroller which drives the memory bus of the prototype.

6.7 Overview and future work

As mentioned in the introduction of this chapter, cost efficiency is one of the driving forces behind the integration of CMOS circuits. Thanks to the ever increasing speed and density of on-chip transistor devices, the power of digital integrated systems soon achieved a critical mass which allowed digital processing to actively participate in a domain which until then had been exclusively reserved for analog circuits: the world of wireless communication. Until that moment, all of the signal processing was performed in the analog domain, with the contribution of the digital back-end being limited to some dedicated mixed-signal blocks such as the PLL or the symbol-to-bit demapper in the rear of the receive chain. The arise of digital signal processing into the field of wireless communication introduced some major changes in the way both analog and digital information is being transferred over the analog wireless channel.

The role of digital signal processing

First of all, signal processing allows to deal with analog information in the same way as is done for digital data. Before being transmitted over the channel, analog information is translated to the digital domain. The advantage of this approach is that digital information can be taken to a higher level of abstraction. Since digital data can be handled as numerical symbols, it becomes very convenient to perform calculations on it. For example, it is possible to add some amount of redundant information to a string of digital data, which can be used later on in the transmission chain to recover the original information in case of transmission errors in the channel. Also, digital information is a lot more flexible to deal with than an analog-valued signal, in this respect that digital bits can be easily buffered and restacked in order to form new *symbols*. These data symbols, in their turn, are converted into a new analog domain representation which is adopted to the oddities and problems that are faced during the transmission of analog signals over the wireless (or wireline) channel.

In **Chapter 2**, it was observed that the underlying mechanisms of error coding are based on the representation of information using *sequences* of symbols. Rather than using individual symbols to encode information bits, multiple bits are encoded at once in a sequence of symbols. Redundant information is embedded in the transmitted symbol stream by only allowing a certain *subset* of allowed sequences out of the total set of all possible symbol sequences. Using a clever subset of allowed sequences, the *distance* between any two possible sequences in the allowed subset can be effectively increased. Of course, the receiver is fully aware of this fact, and will check the received sequence against the current subset of allowed sequences. If a transmission error occurs and some symbols get corrupted during transmission over the channel, the receiver can recover the correct sequence by searching for the allowed sequence with the shortest distance from the received symbol sequence.

It was also found that this *distance* between sequences should be further specified: when data is transferred over the channel in an analog shape, certain errors will occur more frequently than others. For example, it is more likely for a receiver to confuse two neighbouring constellation points of a symbol than two more distant points (either in amplitude or in phase) in the constellation plane. An error correction mechanism which does not take into account the *Euclidean distance* between symbols, will waste much effort and signal energy in adding redundant information to distant constellation points, as a result of which the overall throughput of the system shifts further away from the theoretical throughput capabilities of the channel. This observation has lead to the concept of *modulation-aware coding*, in which the error coding mechanism is aware of the way in which the analog signal is injected in the channel and also knows exactly how the channel affects the transmitted signal in terms of noise and interference.

Joint coding and modulation

In **Section 2.4**, the general idea of modulation-aware error coding was further explored using the example of the 33.6 kbit/s analog voiceband V.34 modem system. The V.34 modem standard employs *Trellis coded modulation* (TCM), which was invented by Gottfried Ungerboeck in 1982 [Ung82]. TCM is a modulation-aware channel coding technique for bandwidth-limited channels. In the TCM scheme, information bits are pushed through the channel at the maximum theoretical throughput rate. At first sight, it might seem impossible to attach redundant information to the transmitted symbol stream, because the channel is already used at the maximum symbol rate. However, Ungerboeck managed to solve this problem by increasing the modulation depth of the system to below the noise floor of the channel. Doing this way, the increased data rate of the transmitter allows to attach redundant information to the transmitted symbol stream, while the rate of effective (unencoded) information does not surpass the theoretical capacity of the channel. Two important ideas were contained in Ungerboeck's way of doing. First, by modulating the signal with an accuracy which is below the noise floor of the channel, the discretizing effects of the symbol-to-bit demapper (and the DAC/ADC) are ruled out of the picture, while only the noise floor of the channel itself remains responsible for the signal-to-noise ratio of the received symbol stream. Secondly, modulating below the noise floor of the channel does not increase the error rate of the system, because only the minimum Euclidean distance between sequences (which is set by the noise floor of the channel) is important to the decoder in the receiver.

The advantage of the Trellis coding process is that a sequence of information bits is mapped onto a sequence of symbols at the output of the receiver, while the information that is contained in a single incoming bit is spread over a sequence of multiple symbols at the output of the system. The net result is that a single error in the received symbol stream affects multiple information bits at the same time, which means that the channel noise is effectively smoothened over multiple information bits. For as long as the bit energy over thermal noise density stays above a certain level ($E_b/N_0 > -1.59$ dB), error-free communication should be possible. The longer the sequence over which the information of a single unencoded bit is spread, the better noise energy can be levelled over multiple bits and the bigger the chances of survival are for the error correction mechanism. However, it should be intuitively clear that the larger the set of possible sequences is, the more difficult it becomes for the error decoder in the receiver to find the sequence with the highest probability of being transmitted. Which, of course, will be reflected in the power consumption profile of the system.

Communication over a wideband radio channel

From the considerations of the system in Chapter 2, it has become clear that an appropriate channel coding and modulation technique is indispensable for the optimal functioning of a communication system. Of course, since the original topic of this text includes the design of a broadband *wireless* radio system, it seems quite logical to shift the center of attention a little, from a wireline to a wideband wireless channel. While the broad outlines and the ideas made for narrowband wireline communication systems still hold, the characteristics of the wideband wireless channel make it everything but fit for use in a communication application. First of all, there is the problem of multipath (**Chapter 4**): the transmitted radio signal has not exactly the intention to travel in a straight line between the transmitter and the receiver. Instead, several reflections of the transmitted symbol stream arrive at the receiver, each with a different time delay. Depending on the ratio between the *delay spread* of the channel and the period of the transmitted symbols, two different things may happen: If the multipath delay spread is shorter than the duration of a symbol, the received streams will start to interfere with each other at the symbol level. This means that several delayed versions of the same symbol arrive at the receive antenna at the same moment, each with a different phase rotation. The result, *frequency-flat fading*, causes intermittent channel outages during which communication becomes impossible. The only feasible way to deal with such a channel is to use an error correction scheme that spreads information over a longer duration in time, which hopefully can bridge the entire period of a link failure. However, this method does not bring any relief under static channel conditions: when both transmitter and receiver are stationary devices in a stable environment, the duration of periods of destructive interference may grow beyond the time-diversity capabilities of the error correction mechanism, with the obvious tragic consequences.

The situation becomes completely different for a broadband communication system. The large spectral footprint of such a radio transmission makes that the channel has a frequency-selective instead of a frequency-flat profile. From a system-level point-of-view this is a good thing, because the diversity of a frequency-selective channel makes it very unlikely that all frequency bands suffer from destructive fading at exactly the same moment. Using the appropriate subchannel loading techniques, it should be possible to pull off a reliable communication link over a frequency-selective channel. Finding out the correct way to inject the information in the wireless channel is not evident, though: from a time domain point-of-view, the frequency dependent channel response is again caused by the delay spread in a multipath environment. This time, due to the wideband character of high-speed communication, the delay spread of the channel may become larger than the symbol period of the transmitter. This implies that delayed versions of the same symbol start to interfere with adjacent symbols, an effect which is known as *intersymbol interference* (ISI). In order to deal with the degrading effects of ISI on the symbol error rate of a transmission, a combination of channel equalization and *orthogonal frequency division multiplexing* (OFDM) is commonly being employed by wireless network devices (e.g. 802.11a/g). In contrast to an equalization filter, which flattens the frequency response of the channel, OFDM splits up the frequency band in multiple parallel communication subchannels. Each of the subchannels in the OFDM system handles a symbol rate which is considerably lower than the delay spread of the channel, which wipes out the problem of ISI. However, the transfer of information remains impossible in those frequency bands which are blanked out by deep fades. The result is an excessive rate of bit errors in certain subchannels of the OFDM system. In the ideal scenario, the controller of the system would avoid transmission over subchannels with an excessive noise level and redistribute the information over other, unaffected subbands. Obviously, this approach requires an active intervention from transmission side, while the receiver must keep the transmitter informed on the current channel conditions. The latter aspect turns out to be a big problem under varying channel conditions: by the time the transmitter gets informed about the transfer characteristic measurement performed at receiver side, the information may already be outdated.

This is one of the main reasons why OFDM-based systems employ bit-interleaved coded modulation (BICM), which distributes information at the bit level over multiple subchannels of the OFDM transmission. The general idea behind this joint coding and modulation technique is to evenly spread the errors over a longer sequence of symbols and as such avoid bursts of erroneous bits being fed to the (Viterbi) decoder in the receiver. The net result of combining OFDM with BICM is that the problem of ISI is in fact transformed into a problem of Gaussian noise, which turns out to be the particular specialty of a traditional forward error coding (FEC) mechanism. However, major opportunities are being missed here to make real advantage of the frequency selectivity of a wideband transmission channel.

Modulation-aware decoding: signal reconstruction

With these thoughts in mind, and from the fact that OFDM is not exactly the most energy efficient modulation technique due its large peak-to-average power ratio, **Chapter 3** introduced a new line of approach. Instead of involving the transmitter in the channel coding process, the core activities which deal with the quirks of the channel are shifted from the transmission side to the receiver end. While the transmitter only employs a very basic modulation scheme (wideband QPSK), it is the full responsibility of the receiver to deal with the changing channel conditions, allowing the system to react very responsive in a dynamic multipath environment with a short *coherence time*. For this purpose, the receiver is not only supported by the up-to-date data on the channel response, but will also take the modulation method of the transmission into account. This consideration has lead to the concept of *modulation-aware decoding*, which is the main topic of **Chapter 3**.

In a modulation-aware radio receiver, the system does not actively attempt to avoid frequency bands that are affected by destructive fading. After all, this would be impossible since a basic modulation technique such as QPSK does not offer the fine-grained control over its spectral footprint such as is the case in the OFDM modulation system. Instead, the receiver tries to recover the original transmitted symbol from the remaining, unaffected frequency bands. Theoretically, this is possible because a single-carrier modulation system (such as QPSK) automatically spreads the information that belongs to a single incoming symbol over the entire spectrum of the transmission band. Exactly how the energy is spread, depends on the surrounding symbols in the transmitted stream.

In the new *interferer suppression and signal reconstruction* (ISSR) technique, introduced in **Section 3.2**, the original symbol sequence is recovered by means of an iterative signal reconstruction technique. ISSR is based on a mixed-domain *particle swarm optimization* (PSO) algorithm, and uses repeated conversions between a time-domain and a frequency-domain representation of the received signal. Frequency bands which suffer from excessive noise exposure are excluded from the game and are subsequently reconstructed using redundant information from the unaffected part of the spectrum. Under ideal circumstances, up to 40% of the frequency band of a transmission can be recovered by the ISSR system, without any performance loss in terms of BER. Apart from the evident issues concerning the dynamic range of the signal, the ISSR system does not differentiate between frequency bands that have been lost as a result of noise or those that are corrupted by a narrowband interferer: corrupted bands are filtered out of the signal in the same way as is done for frequency bands which suffer from deep fades. The ISSR system is not a replacement for the forward error coding subsystem and is not suited to recover problems created by additive white Gaussian noise (AWGN). For this reason, the ISSR decoder should always be used as the inner coder of a standard error scheme, such as a convolutional encoder or a Turbo coder. In the latter case, it was is shown in **Section 3.4** that the combination of ISSR and Turbo coding in a noise free frequency-selective channel performs within 0.4 dB of Shannon's theoretical limit. In this respect, the proposed single-carrier QPSK modulation and its subsequent signal reconstruction technique outperforms OFDM-based modulation systems.

Pulse-based radio: increasing multipath resolvability

In **Section 6.4**, it has been pointed out that, despite the problems caused by ISI, a frequency-selective channel is effectively a necessary precondition for a reliable communication over a wireless link which is affected by Rayleigh fading. All the underlying mechanisms can be reduced to the *multipath resolvability* of the wireless system, which points to the ability of the system to distinguish between several delayed multipath components. It turns out that a high symbol rate is beneficial for the resolvability of a radio transmission, since the energy that belongs to a single symbol is spread over multiple subsequent symbols. It was discussed that this will generate a lot of ISI, but no signal energy is lost or created in this process. This in contrast to *intra-symbol* interference, which is caused by multipath components that arrive within the time frame of a single symbol period. Of course, averaged over a longer period of time, no energy is lost as there is an equal chance on positive or destructive interference, but a reliability problem may emerge under static channel conditions.

This observation is especially true for indoor channels: due to the limited amount of delay spread (typically less than a few hundreds of nanoseconds), a considerable amount of multipath energy arrives within a single symbol period, even for a transmission with a high symbol rate. The result is a flat fading power profile, on top of the frequency-selectivity of the channel. The problem can be mitigated by increasing the symbol rate to even a higher level. For example, the multiband OFDM system employs a symbol rate of more than 500 Ms/s [P8007]. Using such a high symbol rate has a direct impact on the power consumption of the system, since the entire signal chain in the receiver has to support the large bandwidth of the baseband signal. **Chapter 5** tackles this problem and introduces a modulation method which decouples multipath resolvability from the symbol rate: *pulse-based radio*.

Instead of transmitting a continuous-time modulated carrier, the pulse-based radio system employs short pulses. In contrast to a traditional transmission, the spectral footprint of a pulse-based radio is determined by the duration and the shape of the pulses, but not by the symbol rate of the pulses. Nevertheless, it was particularly stressed in **Chapter 5** not to focus on the *pulse*-thing in pulse-based radio. Instead of this, a pulse-based transmitter should considered as a QPSK radio, which has been extended with a transmission gate in order to increase the multipath resolvability. The same principle holds for the receiver, which is a regular coherent QPSK receiver, on top of which a pulse-based extension layer has been implemented in the form of a receive window. The receive window is used to generate (equidistant) receive slots and blocks unwanted noise and interferer power from the receiver if no pulse is expected to arrive.[16] Doing this way blocks a lot of interferer power immediately from the input of the receiver, which is a good thing from a linearity perspective.

The input section of the pulse-based receiver also employs a bandwidth compression technique, which converts the wideband character of the received pulses back into a regular baseband QPSK signal. Since this pulse-to-baseband conversion process occurs early in the signal chain of the receiver, it is successfully prevented that the entire receive chain must deal with the wideband character of the pulses. The remaining part of the receiver deals with the baseband signal as any other QPSK modulated signal. Of course, at this point there is still no benefit of using pulses for the reliability of the radio link. For this purpose (**Section 5.2**), the receiver employs a set of parallel pulse-based receive units, which are dynamically allocated to a receive slot within the time frame between to consecutive pulses.

Diversity combining: interleaved ISSR

Thanks to the resolvability of the short pulses, each receive unit will experience an independent fading (virtual) channel. It is evident that each of the receive units still suffers from a fading channel, since closely separated multipath components still arrive within the duration of the receive slot. However, the combined forces of multiple parallel receive units will boost the reliability of the link (**Section 4.4**). For this purpose, the output of the individual pulse-based QPSK receive units is first processed by the *interleaved* ISSR decoder (**Section 5.3**). In the interleaved version of ISSR, the decoder employs the unaffected bands supplied by different receive units in order to reconstruct the original signal. The result is a much more reliable radio link because – except for the case of interference – chances are very slim that the same frequency band is unavailable at the same moment for different receive units operating in (virtual) channels that have independent fading characteristics.

Finally, the whole concept of pulse-based radio with parallel receive units has an additional advantage, related to the synchronization of the receiver on the received pulse stream (**Section 5.2**): instead of chasing after every individual 'pulse' that may arrive at the antenna, the receiver deliberately ignores the fact that pulses shift in and out of the receive slot allocated to a certain pulse-based unit. This is caused by changes in the arrival time of the pulses, either due to a varying propagation length of a certain multipath component or due to clock offset between transmitter and receiver. Of course, this will result in a rotating constellation diagram in the baseband output of the QPSK receiver, but this issue must be handled at the level of the signal processor in the back-end of the receiver.

[16] The length of a receive slot is longer than the pulse duration to simplify the synchronization process.

Suggestions for continued research

Although Chapter 6 describes the implementation details and measurement results of a pulse-based receive unit, still a lot of work needs to be done before all pieces can be assembled into a ready-to-use system. First of all, the front-end described in Chapter 6 only includes one single receive unit, while several of such units are working in parallel for the pulse-based radio system as described in Chapter 5. This brings some challenging problems to the surface, since the signal power of a single antenna is then split over the input stages of multiple receiver units. If all units share the same physical antenna, a low-noise preamplifier may be required to compensate for the inevitable reduction in signal power available to each receive unit. Remember from Section 6.3 that the LNA is exposed to the full interferer power at the input of the receiver, so this may cause problems in a channel with strong blocker levels.

Furthermore, the pulse-based radio principle, as described in Chapter 5, still requires some additional effort in the area of digital signal processing in the back-end section. The computational costs of the ISSR algorithm in terms of hardware and energy consumption were ignored in the high-level description of Chapter 3, since the exact figure depends on a number of design-time decisions. This includes, for example, the trade-off between the degree of serial computing and parallelism in the DFT-transform of the ISSR algorithm. However, some preliminary calculations show that the cost of ISSR is fairly low, compared to the amount of processing power required by a Viterbi decoder or a Turbo coding system.

A summary of the parameters that were used to support this statement can be found in Table 6.1. During the simulations in Chapter 3, the number of iterations was varied between 4 and 100 ISSR loops. A nominal number of 10 iterations and a vector length of $n = 1{,}024$ symbol samples (Figure 3.6, p. 56) were considered as the indicative numbers for the calculations. The computational cost of ISSR is determined by the number of multiplications, which is the most expensive arithmetic operation. Under the assumption that the efficient fast Fourier transform (FFT) is being used to calculate the DFT and its

Number of ISSR loops :	4...100 (nominal: 10 loops)
Cooley-Tukey FFT ($n = 1{,}024$ symbols) :	$n/2 \cdot \log_2(n) = 5{,}120$ I/Q-mult.
# I/Q-multiplications per ISSR iteration :	$2 \cdot 5{,}120$ approx.
# I/Q-mult. per QPSK sample, per loop :	$2 \cdot 5{,}120/n = 10$
# I/Q-mult. per bit, per loop :	$2 \cdot 5{,}120/n/2 = 5$
# I/Q-mult. per bit for 1 full ISSR run :	20...500 (10 loops: 50)

Table 6.1. Computational complexity of the ISSR algorithm.

inverse (Figure 3.4, p. 53), the approximate number of complex multiplication operations for each ISSR loop is given by Equation (6.4) [Coo93]:

$$\#\,\text{IQ-multiplications/ISSR loop} = 2\,\frac{n}{2} \cdot \log_2(n)\,, \qquad (6.4)$$

where n is the length of the symbol vector at the input of the ISSR loop. For a symbol vector length of $n = 1{,}024$ QPSK samples, this leads to a number of $102 \cdot 10^3$ complex multiplications, for each run of the ISSR algorithm (i.e. 10 iterations). This boils down to a number of 5 complex multiplications/bit per ISSR iteration or 50 complex multiplications for each decoded bit (10 iterations). This value is very low compared to the complexity of a Turbo decoder based on the APP algorithm [Bah74], which requires about 192 multiplications per decoded bit, per iteration [Pin01].

Finally, some additional research may be necessary to address the problem of interfacing the chip with the outside world. The wideband nature of the wideband signals that is so typical to wideband communication systems causes significant reflection and termination problems in a wire-bonded chip. This issue was already brought to the attention of the reader in Figure 6.4, which shows a decreasing trend line in the sensitivity of the receiver for increasing frequencies. Using a flip-chip technique may reduce the degradation caused by the parasitic inductance of the bonding wires. However, extending the microstrip lines further down to the chip level will cost a lot chip area, because of the minimal width of the microstrip traces.

Chapter 7

NONLINEAR LOADED OPEN-LOOP AMPLIFIERS

This section describes a wideband amplifier, which is based on a linearized open-loop topology. The amplifier itself was originally embedded in the pulse-based receiver of Section 7.6, serving as variable gain amplifier (VGA) in the baseband section, but a stand-alone chip version has also been implemented to allow for more extensive characterization. Although the linearized amplifier was specifically intended for use in the signal chain of a wideband receiver, it is equally well suited for application in, for example, the high-speed residue amplifying stages of a pipelined AD converter.

While nonlinearities of the active element are commonly suppressed by using the amplifier in feedback configuration, the lack of excess loop gain in the higher frequency range renders this approach useless for use in high-speed applications. For a detailed explanation on the subject of frequency-dependent distortion suppression, the reader is referred to Appendix 7. In contrast to the feedback-oriented approach, the open-loop amplifier that is described below, relies on a completely different distortion suppression mechanism. The linearized open-loop amplifier can be split in two sections. The first section is the active amplification stage, and provides the useful gain of the amplifier. The second part, the loading stage, is in charge of counteracting the weak nonlinear effects that are introduced by the non-idealities in the amplification stage. It is easy to understand that the mathematical solution to tackle this problem would simply be to reverse the nonlinearities from the first stage. In any real-life implementation however, the options to achieve this goal are limited; the nonlinear transfer characteristics of the amplification stage are caused by the physical properties of the composing transistor devices. As the transistor parameters are fairly incalculable during design time – especially towards higher frequency bands –, measures that try to counteract these non-idealities

W. Vereecken and M. Steyaert, *Ultra-Wideband Pulse-based Radio: Reliable Communication over a Wideband Channel*, Analog Circuits and Signal Processing Series,

using a straightforward approach are fairy ineffective. However, by playing off the same active devices in the amplification stage and the compensation stage against one another, the compensation method introduced further on in this section is inherently immune to variations or modelling errors in the device parameters. Before going deeper into detail, it must be recognized that the most important characteristic of the nonlinear loaded open-loop amplifier discussed in this section is that the voltage gain of one single stage is intentionally kept low, in the range of $0\ldots6$ dB. The exact reason for this decision will become clear later on, but the fact that a low output impedance is combined with an open-loop topology, makes the presented baseband amplifier one of the fastest currently available in a standard CMOS technology.

7.1 Interstage coupling of open-loop transistor stages

Before switching to a more in-depth discussion of the characteristics of the open-loop amplifier topology, this section will first address a major design challenge concerning multistage architectures: on-chip interstage coupling. Interconnecting two analog circuits is not evident, especially when the operating point voltage levels at their interface are different. However, even when the biasing point of one stage lies within the range of voltages accepted by the input of the subsequent stage, the transition remains a tricky operation, as is explained by the following reasoning.

Suppose a differential amplifier is chosen as the basic amplifier module. A first option is then to get rid of the common mode biasing voltages by using capacitors, as at first sight it may seem that they do not interfere with the differential nature of our signal-of-interest. For a single-stage amplifier – a differential LNA for example – this certainly is an acceptable decision. Also, it is a viable solution for a chain of AC-coupled amplification stages, as is the case for the variable gain amplifier found in the IF-stage of a receiver front-end. Things are getting more complicated when the amplifier must handle signal components at near-DC frequencies, and capacitive coupling is not an option. This is, for example, the case in the low-IF and baseband stages of a direct conversion receiver. AC-coupling between two subsequent stages would result in an unmanageable large coupling capacitance which would either consume a big chunk of chip area or which has to be implemented by means of an (expensive) off-chip capacitor.

Consequently, the implementation of a baseband variable gain amplifier is forced in the direction of DC-coupled stages, as is being illustrated by Figure 7.1. While common-mode noise can still be safely ignored since it is suppressed by the common-mode rejection capabilities (CMRR) of each of the amplifiers in the chain, the DC path between consecutive stages introduces a

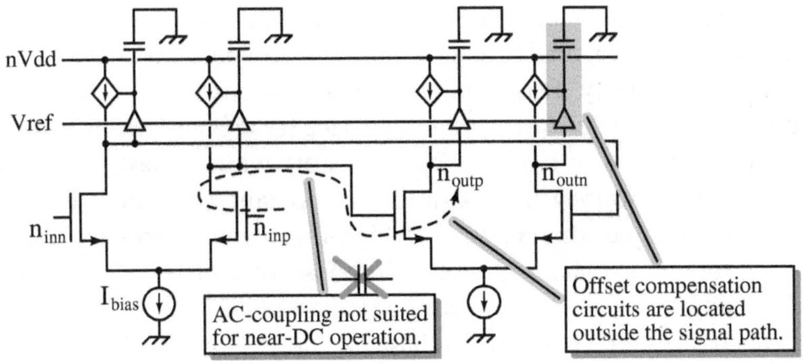

Figure 7.1. For open-loop amplifiers operating at near-DC frequencies, capacitive interstage coupling requires very large (off-chip) coupling capacitors. Differential compensation brings the offset problem outside the signal path, with only minimal consequences for the high-speed performance of the amplifier.

problem caused by one special case of differential 'noise', which is the DC-offset between the outputs of the differential amplifier. Unfortunately, in a DC-coupled chain of amplifiers, the subsequent stage will handle this DC-offset as any other valid input signal. The DC-offset will be amplified, passed to the next stage, amplified again, and so on.

Chances are pretty high that at a certain point further on in the chain, amplifiers will start to clip due to unrestrained offset voltage levels. But even if clipping does not occur for some reason, large offsets exist between the operating point voltages in the two branches of the differential amplifiers. The difference in biasing conditions will also induce an increase of the common-mode to differential conversion gain (more information in Section A.2). In the latter case, the immunity of the amplifier against power supply noise or in-coupling common-mode noise is thus also affected. At this point, it becomes clear that it is essential to introduce common-mode and offset regulating circuits into a DC-coupled multistage amplifier. While the speed of common-mode compensation should be as fast as possible in order to avoid inter-domain signal conversion, the maximum speed of the offset compensation circuit will be determined by the lowest frequency component in the signal of interest. Obviously, the part of the spectrum that falls within the operating speed of the offset compensation circuit will be suppressed. This is especially important for baseband amplifiers, because they require a low cut-off frequency at the lower end of the spectrum.

The reader may correctly argue that the problem of the large interstage coupling capacitances has now become a problem of generating a low-frequency pole in the offset compensation circuit. There is one major difference between the two, however. In the former case, the coupling capacitances lie directly inside the signal path, while in the latter case, the very low time constant necessary for offset regulation can be brought to some point outside of the signal-carrying highway. This is quite an important observation, since it leaves the unenviable analog designer a lot more space for using tricks in order to pull the low-frequency pole as close as possible to DC, without wasting a lot of precious silicon area. Among the possibilities to obtain a decent low-frequency pole are, for example, the creation of high-impedance nodes by employing a cascoded topology, enhanced with gain-boosting if necessary. But also advanced techniques such as the switched capacitor (switch-cap) technique or the even more challenging charge-pump based analog filters may be a possibility.

7.2 Design considerations on the open-loop core stage

The limited voltage gain of the presented open-loop amplifier makes it a perfect candidate for two specific application areas. First, there is the popular pipelined AD-converter. For an in-depth discussion concerning this matter, the reader is referred to more specialized literature, but the most interesting feature about this architecture is that the AD-conversion is spread over several cascaded stages, each resolving 1 or more bits during the digitization of the input signal. In between each of these stages, the analog *residue* signal sample must be amplified before it can be applied to the input of the next stage. The requirements of this interstage amplifier are fairly simple: (1) it must have a certain gain (preferably 6.02 dB), (2) it should be pretty fast and (3) it must exhibit sufficient accuracy, depending on the resolution of the converter.

In most designs, a lot of effort goes into the fine-tuning of the gain and the bandwidth of the feedback-based interstage amplifier. But all too often, the importance of the DC-gain in the interstage amplifier is overestimated. Since the amplifier is used in a sampled system, an accurate low-frequency gain indeed offers good performance at the lower end of the spectrum. But for frequencies closer to the Nyquist bandwidth of the converter, settling time and steady-state error become tightly connected to the excess gain that is still available at that particular frequency. The result is that the effective number of bits of the converter will drop drastically for increasing signal frequencies. Among others, this is one of the main reasons why the figure of merit (FOM) of an AD-converter should always take the effective resolution bandwidth (ERBW) into account, rather than the raw sample rate of the converter.

The second application domain of a high-speed open-loop amplifier is the variable gain pre-amplifier (VGA) in the front-end of a radio receiver. Cascading several low-gain stages is a very efficient way to obtain an input stage with considerable bandwidth and dynamic range. By dynamically inserting or switching off the redundant gain stages, the performance of the baseband-stage can be easily adjusted to a wide range of varying input signal conditions. There are also less obvious advantages of cascading multiple amplifiers: whenever the received signal becomes stronger and less stages are necessary to bring the signal to the required output level, the overall bandwidth of the remaining stages will improve, generating some opportunities to scale the power consumption even more than would be expected at first sight.

Architecture of the open-loop amplifier

The core cell of the open-loop amplifier, without the supporting circuitry to maintain the correct operating point, is shown in Figure 7.2. It is built around a simple differential pair with an active load that consists of a second differential pair in diode configuration. The voltage gain of this configuration is set by the transconductance ratio g_{mi}/g_{mo} between the gain- and the load pair. For maximum gain, the load pair should be omitted. In this case, the output current-to-voltage conversion is taken into account by the parasitic output impedance of the current sources and the gain pair itself.

Of course, the higher the impedance at the output, the lower the bandwidth. When a higher bandwidth is required, the gain of the loading pair can be increased up to the point where the dimensions of both the in- and output pair become equal. The $1/g_m$ impedance of the active load will dominate the output impedance and the voltage gain of the amplifier is (approximately) reduced to

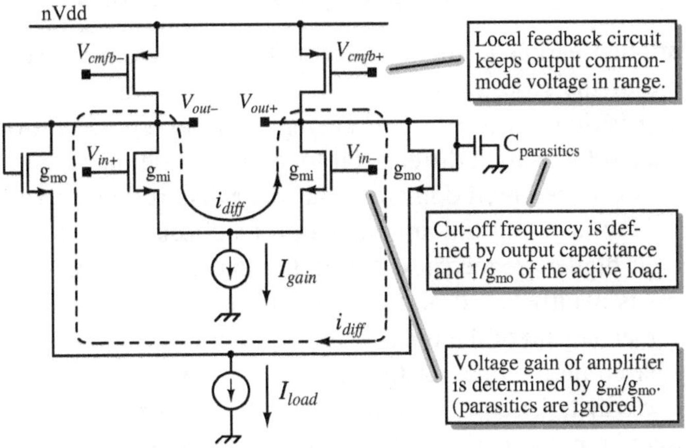

Figure 7.2. Core cell of the open-loop amplifier. The differential output current of the gain stage is directly injected into a second, diode-connected differential pair. Common-mode signal suppression is determined by the parasitic output impedance of the bias current cells feeding both transistor pairs.

unity gain. At this point, the maximum bandwidth of the amplifier is achieved. It is defined by the $1/g_m$ load resistance in parallel with the capacitance of all devices connected to the output terminal: these include the drain capacitances of gain and load transistors, the output capacitance of the current source, the input capacitance of the subsequent stage, the common mode feedback circuitry and last, but not least, some wiring capacitance. In other words, the gain pair will experience a load that is at least $N = 5 \ldots 6$ times larger than its self-loading capacitance, at least for unity gain. Provided that the voltage gain of the amplifier will be in the range of $0 \ldots 6$ dB, a reasonable estimation for the 3 dB bandwidth of a single stage is approximately $f_T/12 = 8$ GHz (6 dB gain, $f_T = 100$ GHz) up to a theoretical maximum of $f_T/5 = 20$ GHz (unity gain, $f_T = 100$ GHz). In practical implementations, the effective bandwidth that is achieved will even be lower than this value.

In order to achieve a better voltage gain, multiple stages of a single core cell are embedded in a chain (The mathematical derivation for the equivalent 3 dB bandwidth of a cascaded amplifier is included in Section A.1). For example, suppose an eight-stage amplifier in which the bandwidth of each core cell is $\text{BW}_{\text{3dB, core}} = 8$ GHz. From Formula (A.14), it follows that the overall 3 dB bandwidth of the eight-stage amplifier is three times less than the bandwidth of its composing stages. In worst case, for an $f_T = 100$ GHz technology, an experienced designer should end up with a BW_{3dB} around 2.5 GHz, which is still remarkably better than what can be achieved in a closed-loop approach.

7.3 Improving linearity using a nonlinear load

It was mentioned earlier in this chapter that the nonlinear transconductance characteristic of the gain transistors in a differential pair can be suppressed by employing various strategies. For example, even-order distortion is effectively eliminated by using a differential topology, while the signal-of-interest is situated in the differential domain. Odd-order distortion on the other hand, is suppressed by inserting the amplifier in a feedback structure, but this approach fails at higher frequencies. While the distortion suppression of feedback systems is mainly based on the trade-off between gain and linearity, the suppression mechanism of the open-loop amplifier discussed in this chapter is completely based on the linearization of the transconductance gain by the inverse I-to-V characteristic of a nonlinear active load. At the lower end of the frequency spectrum, the performance of the feedback topology is unrivaled if compared to the open-loop linearization technique. But in terms of suppression bandwidth it is undoubtedly the latter one that beats the lot.

Furthermore, it has already been discussed that the voltage gain of the nonlinear loaded amplifier from Figure 7.2 can be controlled by adjusting the transconductance ratio g_{mi}/g_{mo} between the gain and the loading transistors.

A designer has several options to achieve the required transconductance ratio. During design time, the transconductance of each pair can be altered by choosing the appropriate W/L dimensions for each transistor. It is also possible to dynamically adjust the tail current at runtime: a lower tail current of the diode-connected load pair results in a higher output impedance and thus an increased voltage gain. How tempting it may look to have stepless control over the gain stages, the tail current will not be used as an extra degree of freedom in the last four to five stages of the amplifier chain: it is precisely the tail current ratio which plays a crucial role in the optimization of linearity of the open-loop amplifier.

This is clarified by the following intuitive analysis. Suppose that both the gain- and the loading pair are perfectly symmetrical, which also applies to their tail currents. The voltage gain of this setup is very close to unity.[1] The nonlinearities of the transconductance gain stage are suppressed thanks to the inverse I-to-V characteristic of the active load. Now, in order to increase the gain, the W/L dimensions of the loading pair are slightly decreased, *without* altering the tail current of the loading pair. As a result of the reduced $1/g_{mo}$ impedance of the diode-connected pair, the voltage gain of the amplifier will increase with same factor.

At this point though, the proposed distortion cancellation mechanism needs some extra attention, because the signal swing applied to the active load has become larger than the signal swing over the input terminals of the amplifier. At first sight it may seem that, due to the larger signal swing at the output, the increased third-order nonlinearity of the active load will cause problems. However, it is important to remember that, although the W/L dimensions of the load were decreased, the tail current was kept constant. As a result, the overdrive voltage ($V_{GS} - V_T$) of the active load will increase automatically, which in turn results in an improved linearity [San99]. Summarizing, the increase of distortion resulting due to the larger signal swing at the output is counteracted by a larger overdrive voltage of the active load.

The outcome is that, after some fine-tuning of the tail currents, third-order distortion is still effectively suppressed by the inverse I-to-V characteristic of the active load. Remark that, even though the proposed linearization technique is not based on any form of feedback, the quality of the linearization mechanism still deteriorates near the 3 dB cut-off frequency of the core amplifiers: at higher frequencies, a substantial portion of the output current of the gain pair is absorbed in the parasitic load capacitance at the output terminals. It follows that the neutralization of the imperfections of the transconductance become

[1] Provided that the drain resistance of the transistors is sufficiently high so that the $1/g_m$ resistance of the diode-connected load dominates the output impedance.

less accurate and distortion levels will rise. Nonetheless, since the core cell amplifiers are at least an order of magnitude faster than the cut-off frequency of a higher-order feedback system, a remarkable good distortion suppression is still achieved at frequencies where most closed-loop amplifiers are only a shadow of their former selves.

7.4 Distortion analysis of the nonlinear loaded stage

In the previous section, it was intuitively demonstrated that third-order distortion in the open-loop amplifier topology of Figure 7.2 could be suppressed thanks to the inverse characteristics of the diode-connected differential load. It was also cleverly observed that the signal swing at the output of the average amplifier is larger than the swing at its input. Thus, in order to remain good distortion suppression performance for gains larger than unity, the transconductance characteristic of the loading pair had to be linearized by means of a larger overdrive voltage. This goal was achieved by maintaining the same tail current for both the gain- and loading pair. In this section, it is mathematically shown that for the simple quadratic transistor model, third-order distortion completely disappears when the tail current of the gain pair is chosen equal to the tail current of the load. For the more general case, where the transconductance of the transistor deviates from the idealized quadratic model, either towards a more linear or a higher order characteristic, a mathematical proof would not bring additional insight in the matter due to the high complexity, but the insights brought by the calculations hereunder should remain valid.

In the mathematical solution below, the quadratic g_m characteristic from [Lak94] has been used. The large signal model of an NMOS in strong inversion depends on some physical parameters represented by K_n', its dimensions W (effective gate width) and L (channel length) and the overdrive voltage $V_{GS} - V_T$ in the following way (7.1):

$$\begin{aligned} I &= K_n' \frac{W}{L} (V_{GS} - V_T)^2 \\ &= \frac{\beta}{2} (V_{GS} - V_T)^2 \end{aligned} \tag{7.1}$$

Our first target is to find an expression for the transconductance when the transistors are embedded in a differential pair. It is clear that due to the symmetrical setup, even-order components will disappear from the transfer characteristic. In case of an ideal current source I_{bias} forcing a constant bias current through the transistor pair, the following equations hold (7.2):

$$\begin{aligned} i_{od} &= i_{noutp} - i_{noutn} \quad \text{(differential current)} \\ I_{bias} &= i_{noutp} + i_{noutn} \quad \text{(common mode current)} \end{aligned} \tag{7.2}$$

When the first order model (7.1) for a MOS transistor in saturation is plugged in, the relationship between the differential input voltage ($v_{id} = v_{ninp} - v_{ninn}$) and the differential output current (i_{od}) is given by (7.3):

$$\begin{aligned} v_{id} &= v_{GS1} - v_{GS2} \,, \\ \text{with } v_{GS1} &= V_T + \sqrt{\frac{i_{noutp}}{K'_n \cdot (W/L)}} \\ \text{and } v_{GS2} &= V_T + \sqrt{\frac{i_{noutn}}{K'_n \cdot (W/L)}} \end{aligned} \tag{7.3}$$

From Equation (7.2) an expression for both i_{noutp} and i_{noutn} as function of I_{bias} and i_{od} can be extracted. When substituted into the flattened relationship between the differential input voltage and the differential output current (7.3), the following result may be obtained[2] (7.4):

$$v_{id} \cdot \sqrt{\frac{2K'_n(W/L)}{I_B}} = \sqrt{1 + \frac{i_{od}}{I_B}} - \sqrt{1 - \frac{i_{od}}{I_B}} \tag{7.4}$$

This equation describes the differential input voltage as a function of the output current, while the inverse function represents the transconductance of the differential pair. The Taylor series expansion of the inverse function of (7.4) can be obtained as follows (7.5):

$$\begin{aligned} \text{suppose } x &= v_{id} \cdot \sqrt{\frac{2K'_n(W/L)}{I_{bias}}} \quad \text{and} \quad y = \frac{i_{od}}{I_{bias}} \\ \text{then } x &= \sqrt{1+y} - \sqrt{1-y} \\ y &= x \cdot \sqrt{1 - \frac{x^2}{4}} \\ y &\approx x - \frac{1}{8}x^3 - \dots \\ \frac{i_{od}}{I_{bias}} &\approx v_{id} \cdot \sqrt{\frac{2K'_n(W/L)}{I_{bias}}} - \frac{1}{8}\left(v_{id} \cdot \sqrt{\frac{2K'_n(W/L)}{I_{bias}}}\right)^3 - \dots \end{aligned} \tag{7.5}$$

As expected, no second order distortion components are present in Formula (7.5). Note that this conclusion is valid only for a sufficiently high output impedance of the tail current source of the differential pair. If for some reason the parasitic capacitance at the common node is too high, or there is a

[2] Under the assumption that no calculation errors were made by the author.

matching problem in the gain pair, second order components will emerge at higher frequencies. The *negative* third-order term in (7.5) represents the large-signal gain compression and is also related to the third-order intermodulation distortion IM3 in the frequency domain (more information in Section A.1). At this point, when the differential pair has a resistive load R_{load}, these third-order terms are directly translated to the voltage domain by $v_{\text{out, diff}} = i_{\text{od}} \cdot R_{\text{load}}$. In order to counteract the third-order distortion component, the differential pair has been loaded with a second pair, which has the same differential voltage-current conversion characteristic (7.6):

$$\begin{aligned} i_{\text{od, gain}} &= v_{\text{id}} \cdot \sqrt{\beta_{\text{gain}} \cdot I_{\text{bias}}} - \frac{1}{8} \cdot v_{\text{id}}^3 \cdot \sqrt{\beta_{\text{gain}}^3 / I_{\text{bias}}} \\ i_{\text{od, load}} &= v_{\text{od}} \cdot \sqrt{\beta_{\text{load}} \cdot I_{\text{load}}} - \frac{1}{8} \cdot v_{\text{od}}^3 \cdot \sqrt{\beta_{\text{load}}^3 / I_{\text{load}}} \end{aligned} \tag{7.6}$$

In this formula, v_{id} represents the differential voltage at the input of the amplifier, while v_{od} is the voltage across the output terminals of the amplifier. If the impedance of the load is dominating over all other parasitic impedances at the output node, all differential current from the input pair will flow through the loading pair. In first order, the output voltage of the amplifier can thus be approximated by $v_{\text{od}} = \text{gain} \cdot v_{\text{id}}$ and the following equation holds (7.7):

$$\begin{aligned} & v_{\text{id}} \cdot \sqrt{\beta_{\text{gain}} \cdot I_{\text{bias}}} - \frac{1}{8} \cdot v_{\text{id}}^3 \cdot \sqrt{\beta_{\text{gain}}^3 / I_{\text{bias}}} \\ = \; & v_{\text{id}} \cdot \text{gain} \cdot \sqrt{\beta_{\text{load}} \cdot I_{\text{load}}} - \frac{1}{8} \cdot (v_{\text{id}} \cdot \text{gain})^3 \cdot \sqrt{\beta_{\text{load}}^3 / I_{\text{load}}} \end{aligned} \tag{7.7}$$

By comparing the first order terms of Equation (7.7), the expression for the linear gain of the amplifier can easily be found. Comparing the third-order terms and substituting the expression for the first-order gain leads to the following precondition for a linear voltage gain between in- and output (7.8):

$$\begin{aligned} \text{gain} &= \sqrt{\frac{\beta_{\text{gain}} \cdot I_{\text{bias}}}{\beta_{\text{load}} \cdot I_{\text{load}}}} && \text{from the first-order terms of (7.7)} \\ \sqrt{\frac{\beta_{\text{gain}}^3}{I_{\text{bias}}}} &= \text{gain}^3 \cdot \sqrt{\frac{\beta_{\text{load}}^3}{I_{\text{load}}}} && \text{from the third-order terms of (7.7)} \\ & \Downarrow \\ I_{\text{bias}} &= I_{\text{load}} && \text{precondition for linear voltage gain (7.8)} \end{aligned}$$

For a differential pair with MOS transistors in operated in the saturation region, the large-signal third-order distortion component can thus be avoided by choosing the same tail current for both gain pair and load pair. The voltage gain can be chosen independently from this requirement, and is determined by the ratio $\beta_{\text{gain}}/\beta_{\text{load}}$. One has to keep in mind that the overdrive voltage $V_{\text{GS}} - V_{\text{T}}$ should be chosen large enough so that the transistors remain in the strong inversion region during the entire period of the in- and output signal. Of course, the exact mathematical formulation from above is only valid for as long as the CMOS transistors follow a perfect quadratic model. When their characteristic deviates from this idealized model, finding a closed-form expression for the tail current ratio becomes a really challenging (and unnecessary) job.

Besides this, it is again brought to the attention of the reader that the parasitic output impedance of devices connected to the output node is going to divert some of the output current, thereby making the simplified expression from above quite pointless for the purpose of finding the correct current ratio. Fortunately, apart from that it may be an interesting experiment from a mathematician's point of view, finding the universal expression is not the first concern for the design engineer: as will be explained later on, there are more reliable methods that can provide the exact ratio between the tail currents. What is important to realize, is that the quadratic approximation is of greater value to the design engineer as it can provide a rather intuitive insight in the sensitivity of the distortion cancellation technique towards variations in the design process. It is, for example, far more interesting to understand the effect of mismatch on the performance of distortion cancellation. The impact of offset and mismatch on the nonlinear loaded amplifier will be discussed in the next section.

7.5 Sensitivity analysis of the open-loop amplifier

In the previous section it was discussed that third-order distortion in a nonlinear loaded amplifier can be suppressed for a correct ratio between the tail currents of the gain- and loading pair. For the idealized quadratic model of a transistor in saturation, it was observed that both tail currents should be chosen to be equal. However, in any practical implementation of the amplifier, there will always be a discrepancy between the intended operating point and the effective operational settings. In the previous discussions it was silently assumed that the transistors of the differential pair are perfectly equal. Differences in oxide thickness, substrate doping and mobility however will cause matching errors between the MOS transistors. This will be reflected on important model parameters such as a variation in the threshold voltage (ΔV_T) and a relative offset in the transconductance gain ($\Delta\beta/\beta$) between the transistors [Pel89] (7.9):

$$\begin{aligned} \sigma_{\Delta V_T} &= \frac{\sigma_{0,\Delta V_t}}{\sqrt{WL}} \\ \sigma_{\Delta\beta/\beta} &= \frac{\sigma_{0,\Delta\beta/\beta}}{\sqrt{WL}} + \text{constant} \end{aligned} \tag{7.9}$$

As a result of the difference between the transistor parameters, third-order distortion components will reappear at the output of the amplifier. It is now interesting to predict what can be expected in terms of linearity performance, when some realistic mismatch values are taken into account. In order to have some reference data, first of all the effect of mismatch on a simple resistive loaded differential pair is recapitulated. Then, without going into the gory details of the sensitivity calculations, the mismatch results of the nonlinear loaded amplifier are introduced, followed by a quantitative evaluation of the results.

Mismatch versus distortion in a resistively loaded pair

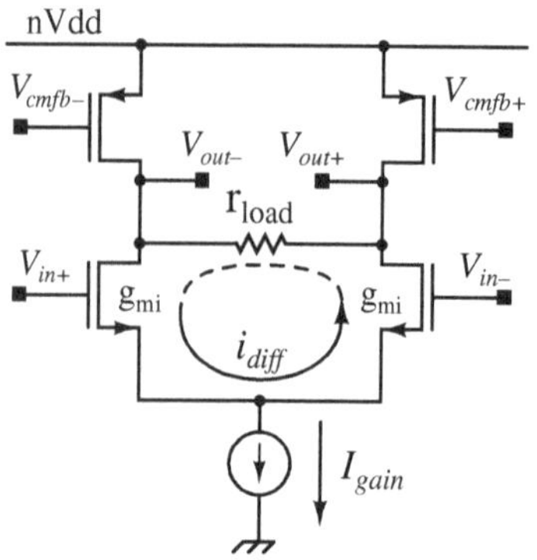

Mismatch in a resistive differential pair causes a DC offset in the current at the output. At first sight this may not seem a very important issue, but the difference in the operating point between the branches of the pair deteriorates the suppression of the common-mode to differential conversion gain. Moreover, first-order mixing products between the differential input and common mode noise will upconvert this noise to the frequency band of interest. Apart from offset, mismatch in a resistive loaded differential pair will cause second-order distortion to reappear, in addition to the existing third-order components. The offset and second-order distortion expressions for the resistive loaded pair are given by (7.10):

$$\begin{aligned} \frac{i_{\text{offset,DC}}}{I_{\text{bias}}} &= \frac{\Delta V_{\text{T}}}{V_{\text{GS}} - V_{\text{T}}} \\ \text{HD}_2 &= \frac{3}{16} \cdot \frac{v_{\text{idp}}}{V_{\text{GS}} - V_{\text{T}}} \underbrace{\left(\frac{\Delta V_{\text{T}}}{V_{\text{GS}} - V_{\text{T}}} + \frac{1}{2} \frac{\Delta \beta}{\beta} \right)}_{\text{additional reduction*}}, \end{aligned} \tag{7.10}$$

where v_{idp} is the peak amplitude of the differential AC-signal v_{id} over the input terminals of the amplifier. Typical values for the standard deviation of the matching parameters of an NMOS device in a 0.13 µm process are $\sigma_{0,\Delta V_t} = 3\,\text{mV} \cdot \mu\text{m}$ and $\sigma_{0,\Delta\beta/\beta} = 0.01 \cdot \mu\text{m}$. For example, suppose that an uncertainty interval of 3σ is taken into account. The yield in this confidence interval is thus 99.7%. Also suppose that the area of the transistors is $WL = 1\,\mu\text{m}^2$ and the overdrive voltage ($V_{\text{GS}} - V_{\text{T}}$) is 130 mV. It follows that, compared to a single-ended transistor amplifier, an additional HD_2 reduction (*7.10) of at least 20 dB is guaranteed.

The influence of mismatch on third-order harmonic distortion HD3 can be neglected. The reason for this is that the third-order harmonics already are a second-order effect: they are the result from the mix-product of nonlinearities originating from (1) the second-order transconductance characteristic and (2) variations of the voltage at the common source node of the differential pair. The same formula as for the ideal differential pair can thus be retained, which is reproduced below for the sake of completeness (7.11):

$$\mathrm{HD}_3 = \frac{1}{32}\left(\frac{v_{\mathrm{idp}}}{V_{GS} - V_t}\right)^2 \tag{7.11}$$

Mismatch and distortion in a nonlinear loaded pair

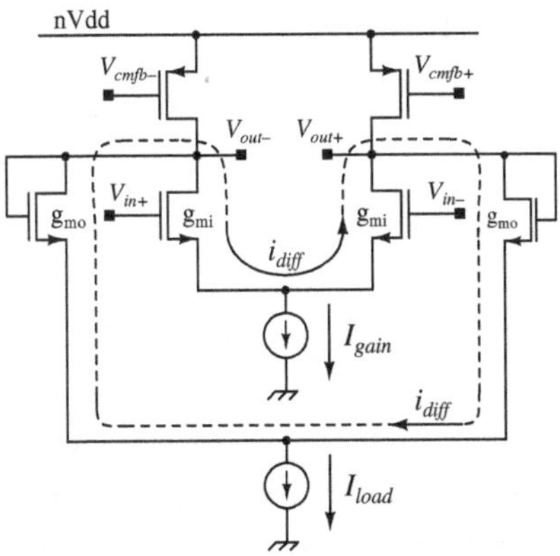

The ideal nonlinear loaded differential pair in Figure 7.2 exhibits no second- or third-order distortion. Similar to the case of second-order distortion in the resistive loaded pair, mismatch will shift the settings for maximum distortion suppression away from the optimal biasing point. It was mentioned in the distortion analysis of the ideal nonlinear loaded pair that mismatch between the transconductance factor β between the transistors of the gain pair (g_{mi}) and the loading pair g_{mo} will change the voltage gain of the amplifier, but there is no impact on distortion as long as the tail currents are the same. However, a mismatch between the mutual parameters within each of the pairs, or an offset in the correct tail current ratio will produce second- and third-order distortion components.

Just like in the case of the resistive loaded pair, it is important to gain some insight in the relationship between the mismatch parameters of a certain technology and the performance that can be expected from the linearized open-loop amplifier. Again, the actual derivation of the mismatch formulas is omitted and only the results are being shown below. The DC-offset voltage of the nonlinear loaded amplifier is given by (7.12):

$$\begin{aligned} v_{\mathrm{offset,DC}} &= \mathrm{gain} \cdot \Delta V_{\mathrm{T,gain}} + \Delta V_{\mathrm{T,load}} \\ &\quad + \frac{1}{2} \cdot (V_{\mathrm{GS2}} - V_{\mathrm{T}}) \cdot \left[\frac{\Delta\beta_{\mathrm{gain}}}{\beta_{\mathrm{gain}}} + \frac{\Delta\beta_{\mathrm{load}}}{\beta_{\mathrm{load}}}\right], \end{aligned} \tag{7.12}$$

where $\Delta V_{\mathrm{T,gain}}$ is the threshold offset between the transistors of the gain pair and $\Delta V_{\mathrm{T,load}}$ is the threshold offset for the loading pair, while $\Delta\beta_{\mathrm{gain}}/\beta_{\mathrm{gain}}$ and

$\Delta\beta_{\text{load}}/\beta_{\text{load}}$ represent the transconductance mismatch of the input (gain) and the output (load) transistor pair. The first part of this expression can be validated quite easily: offset in the input pair threshold voltage ($\Delta V_{\text{T,gain}}$) can be modelled as a fixed offset voltage applied to the differential input of the amplifier. Finding the corresponding offset at the output is then just a matter of scaling the virtual offset voltage at the input with the voltage gain of the amplifier. Offset between the threshold voltages in the output pair is directly visible at the output. The second part of (7.12) is somewhat counter-intuitive, in this respect that a larger overdrive voltage ($V_{\text{GS2}} - V_{\text{T}}$) seems to result in a larger sensitivity towards errors in the transconductance factor β. However, this result is quite logical: for a fixed bias current, a larger overdrive voltage ($V_{\text{GS2}} - V_{\text{T}}$) of the diode-connected output pair corresponds to a increased load impedance ($1/g_{\text{m, load}}$). A current-mismatch between the branches of the input (or output) pair is thus translated in a larger offset voltage at the output.

The second-order harmonic distortion ratio of the nonlinear loaded amplifier can be approximated by (7.13):

$$\text{HD}_2 = \frac{3}{16}\frac{v_{\text{idp}}}{V_{\text{GS1}} - V_{\text{T}}} \cdot \frac{\Delta I}{I} \cdot \left(\frac{2 \cdot \Delta V_t}{V_{\text{GS1}} - V_{\text{T}}} + \frac{\Delta\beta_1}{\beta_1}\right) \tag{7.13}$$

From Formula (7.13), it is clear that the HD_2 is the product of two small values: (1) current mismatch and (2) offset of the threshold voltage. When the tail currents of the gain- and load pair are chosen to be equally, second-order distortion should not be of any significance. However, in the real implementation of the amplifier, a small portion of the output current will flow through the finite output impedance of the transistors connected to the output node. This is even more true for increasing frequencies, at which the impedance of the parasitic output capacitances becomes a significant portion of the output impedance. As a consequence, a more realistic distortion model would take into account that a non-negligible portion of the total differential output current is injected into a linear impedance. This fraction of the output signal should then be characterized by the mismatch expressions of the resistively loaded pair (7.10).

Continuing in the same vein, the reader must keep in mind that third-order distortion partly originates from the fact that a fraction of the output current does not flow through the loading pair. Nevertheless, for the part of the signal that does flow through the diode-connected pair, the following expression holds (7.14):

$$\begin{aligned}\text{HD}_3 &= \frac{1}{32}\left(\frac{v_{\text{idp}}}{V_{\text{GS1}} - V_{\text{T}}}\right)^2 \cdot \\ &\left[2\frac{\Delta I}{I} + 3\left(\frac{\Delta V_{\text{T}}}{V_{\text{GS1}} - V_{\text{T}}}\right)^2 + \frac{3}{4}\left(\frac{\Delta\beta_{\text{gain}}}{\beta_{\text{gain}}}\right)^2\right]\end{aligned} \tag{7.14}$$

It does not make sense to put all effort in calculating the correct tail current ratio or to minimize mismatch between transistor parameters if a significant portion of the output signal is flowing to a parasitic output impedance. The attentive reader may suggest that it may be possible to counteract this additional source of distortion by changing the tail current ratio to a new optimum in one way or another. But it is important to recognize that the resistive character of the loading pair can never compensate for a parasitic capacitor, since the load of these two impedances is 90° out of phase. On the other hand, it is indeed possible to employ a tunable tail current to neutralize an offset in the threshold voltage or the transconductance factor. This could be exploited, for example, during the calibration process of the amplifier. If the amplifier is embedded in a digital receiver, a simple in-circuit two-tone test could be used to discover the optimum tail current settings. After all, it is very convenient for a digital receiver to monitor those frequencies where third-order intermodulation products are expected and then adjust the tail current ratio for optimum linearity performance.

7.6 Implementation of a linearized open-loop amplifier

This section guides the reader through the actual design process of an open-loop linearized baseband amplifier, which has been implemented in a standard 0.13 μm CMOS process. The basic architecture of the amplifier is built around an eight-stage cascade of nonlinear loaded core amplifier cells. The amplifier shows a gain of 30 dB in the broadband frequency range of 20 to 850 MHz. Thanks to the active diode-connected loads that partially compensate the nonlinear transconductance characteristics of the gain transistors, an output-referred OIP_3 of better than 13 dBm could be achieved over the complete frequency band. The chip occupies an area of $0.75 \times 0.75\,mm^2$ and draws a total power of 56 mW from a single 1.2 V power supply.

Architectural reflections on the open-loop amplifier

One of the very first requirements of the amplifier is being able to directly drive the input impedance of external measurement equipment. Doing this way, the process of measuring and characterizing the amplifier becomes much more convenient. Also, the influence of other building blocks (mixer, AD-converters) on the linearity performance is completely ruled out from the equation. In order to combine this requirement a considerable 3 dB bandwidth, the architecture of the amplifier is based on a tapered structure. Just like in the case of high-speed digital buffers, the dimensions of subsequent amplifiers are gradually scaled up towards the final buffer stage.

BANDWIDTH VERSUS LATENCY

There is one noteworthy difference between the analog multistage buffer and its digital counterpart though. In the former case, high bandwidth is obviously one of the major design goals. In the latter case, rather low latency instead of a high bandwidth is preferred. For minimum latency, it can be shown [Com96] that each subsequent stage of the digital buffer must be upscaled with a factor e (Euler's number).

While e is the optimum scaling factor for a minimum-delay digital buffer, concluding that the same magic e-number also holds for the tapered open-loop amplifier is a bridge too far for the self-respecting analog designer. First of all, the open-loop nature of a baseband amplifier does not require that phase-delay becomes the most important design criterion: the variables in the design space of an analog amplifier include power consumption, chip area, linearity, voltage swing, noise figure, maximum output load, gain and bandwidth.[3] The following paragraphs describe the design strategy that was followed during the implementation of the eight-stage amplifier. Also, the correlation between several system parameters will be explained.

The upscaling factor of the tapered amplifier

First of all, the primary goal during the design of this amplifier was maximum bandwidth. This in contrast to a commercial design, where the target would rather be maximum power efficiency. Or minimum design effort. The bandwidth is then chosen in order to meet the minimal requirements of the target application. Starting from the first stage of the amplifier, each of the subsequent stages is upscaled with a factor x and directly drives a small resistive or large capacitive load. In the design presented here (Figure 7.3), the final stage drives the 50 Ω impedance of a vector network analyzer. If impedance matching between the last stage and the measurement equipment is required, the transconductance of the loading pair in last stage must be 20 mS. As a comparison, the output impedance ($1/g_{\text{mo}}$) of the first, smallest amplifier in the design is 900 Ω. Starting from the output impedance of the first stage, the target impedance in the last stage ($R_{\text{load}} = 50\,\Omega$) and the number of stages (n), the correct scaling factor is given by (7.15):

$$x = \frac{1}{\sqrt[n]{g_{\text{mo}} R_{\text{load}}}} \qquad (7.15)$$

[3]Gain-bandwidth (GBW) is not very meaningful in the figure-of-merit of a multistage amplifier.

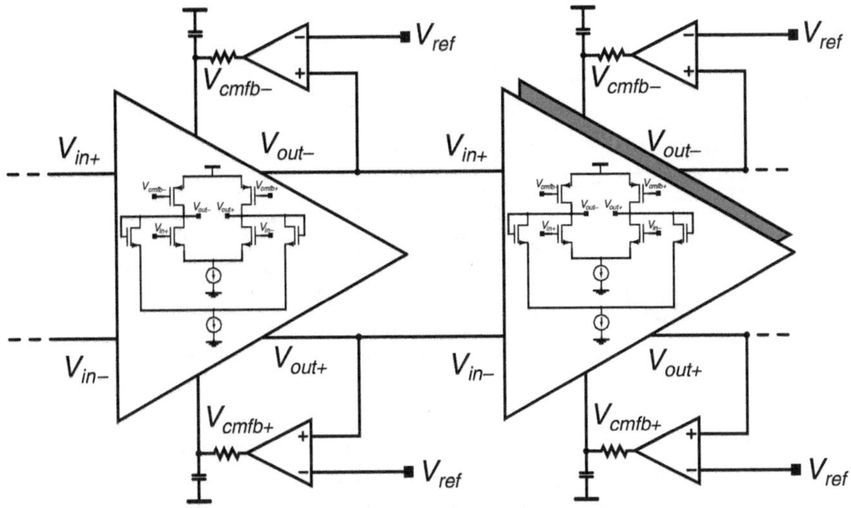

Figure 7.3. Schematic of two cascaded core cells in the amplifier chain. Each section of the amplifier is upscaled with a factor x Equation (7.20) in order to drive a 50 Ω load with a maximum bandwidth. Note that each of the core amplifier cells has its own dedicated offset compensation circuit. A more area-efficient solution would perform the offset compensation over several core cells at once, at the cost of some distortion performance.

Before the upscaling factor x can be determined, one more missing parameters must be determined: the optimal number of gain stages (n) in the chain. At this moment, things will get a little bit more complicated since several interrelated design parameters play a role in our search for the optimum number of stages. The following reasoning clearly shows the counteracting forces that play in this optimization problem. First of all, the 3 dB bandwidth of the complete chain is defined both by the bandwidth of one single stage and the number of stages. A reduction in the number of stages will require a larger upscaling factor between two stages. Which in his turn results in an increased load at the output of each stage. The bandwidth of a single stage decreases and so does the bandwidth of the complete amplifier. However, one parameter has not been taken into account yet: the total voltage gain (A_v) of the amplifier. If A_v is chosen as fixed, independent design parameter, the voltage gain of each single stage depends on the number of stages in the chain. The reader should remember from Section 7.2, that the gain of the open-loop stage was controlled by the $1/g_{mo}$ output impedance of the diode-connected load transistors. As a consequence, changing the gain of the open-loop stage inevitably also changes its bandwidth. Which also reflects on the overall bandwidth of the chain. A rough sketch of all interrelated parameters in the optimization problem can be found in Figure 7.4.

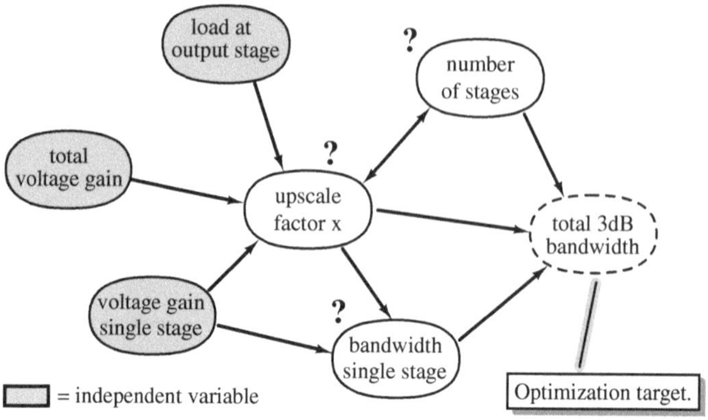

Figure 7.4. Interrelated parameters in the optimization problem of the open-loop amplifier. Things get complicated because a change the upscale factor x triggers a cascade of changes in parameters that are necessary to determine the upscale factor.

In order to solve the previous optimization problem, a simplified model of the amplifier and its properties was used. The target variable that has to be optimized is the bandwidth of the n-stage amplifier, while the only independent variable is the number of stages. Doing this way, the optimization problem becomes relatively simple since all other variables, such as the single-stage gain, the upscaling factor and the bandwidth can be easily deduced from the (discrete) number of gain stages and the target voltage gain. The search for maximum bandwidth is then reduced to a single sweep over the range of the acceptable numbers of stages.

The following examples will use the values that were used in the actual design of the amplifier in a 0.13 µm, 1.2 V CMOS technology. The total required gain of the baseband amplifier was set to 30 dBV. For reasons that will become clear during the layout phase later on, each of the n stages provides an identical voltage gain, defined by (7.16):

$$A_{\text{v, single stage}} = \sqrt[n]{A_{\text{v, total}}} \tag{7.16}$$

The bandwidth of one open-loop stage in the chain is determined by a pole which is defined by the output resistance ($1/g_{\text{mo}}$) of the loading pair and the parasitic load of all devices connected to the output. From some preliminary simulation tests, an approximate value of the most important parasitic capacitors was obtained. In each case, the value the capacitors was referred to C_{in}, which is the gate-source capacitance of the input pair. The value of all capacitors was then combined in the following expression for the capacitive load at the output terminal of the open-loop stage (7.17):

$$C_{\text{out}} = kC_{\text{in}} + \underbrace{\frac{C_{\text{in}}}{A_{\text{v, single stage}}}}_{\text{loading pair}} + \underbrace{xC_{\text{in}}}_{\text{next stage}}, \tag{7.17}$$

where the fairly constant factor k ($\approx$3) embodies the combined parasitic capacitance of the PMOS bias transistors (a), the offset compensation circuitry (b) and some interconnect wiring (c). The gate-source capacitance of the diode-connected pair is equal to that of the gain pair for unity voltage gain. Higher gains are achieved by a proportional reduction of the width of the loading pair. This is taken into account by the second term. The last section in Formula (7.17) denotes the input capacitance of the subsequent gain stage, which is upscaled with a factor x. The approximate bandwidth of the nonlinear loaded open-loop stage, when it is embedded in a tapered structure is given by (7.18):

$$\omega_{\text{3dB, single stage}} = \frac{1}{r_{\text{load}}C_{\text{out}}} = g_{\text{mo}} \cdot \frac{1}{C_{\text{in}}\left[k + \frac{1}{A_{\text{v, single stage}}} + x\right]}$$

$$\begin{aligned}\omega_{\text{3dB, single stage}} &= \frac{g_{\text{mi}}}{A_{\text{v, single stage}}} \cdot \frac{1}{C_{\text{in}}\left[k + \frac{1}{A_{\text{v, single stage}}} + x\right]} \\ &= \underbrace{\frac{g_{\text{mi}}}{C_{\text{in}}}}_{f_{T,\text{ input}}} \cdot \frac{1}{1 + A_{\text{v, single stage}}\,[k + x]},\end{aligned} \tag{7.18}$$

where g_{mi} and C_{in} are the transconductance and the gate-source input capacitance of the first gain stage. It follows from the last expression in (7.18) that the bandwidth of the amplifier can be expressed as a function of f_T, the cut-off frequency of the input transistors. Remark that this frequency is not necessarily equal to the maximum achievable f_T in a certain technology: the nonlinear loaded amplifier achieves a better distortion suppression for a higher output resistance of the transistors: increasing the channel length of the input pair increases the channel-induced output resistance of the gain pair. For example, in the design presented in this section, the physical length of the input transistors was increased from 0.12 to 0.24 µm. Doing so, the parasitic output impedance r_o of the non-ideal transconductance is increased with a factor 3. This is, of course, at the expense of the cut-off frequency, which dropped from 100 GHz

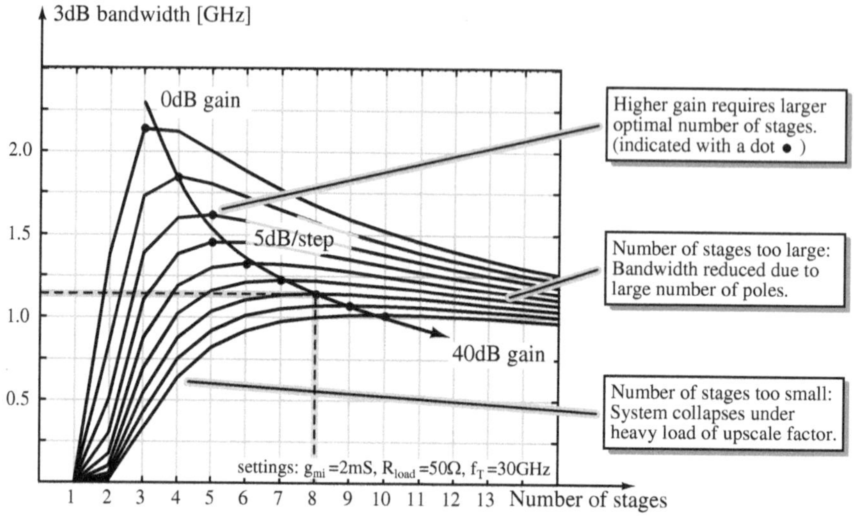

Figure 7.5. Theoretical bandwidth of the open-loop amplifier versus the number of stages (Formula 7.19), simulated for different gain settings (sweep from 0 to 40 dB). It was supposed that the f_T is 30 GHz, taking the load of the transistors into account.

to less than 35 GHz. Finally, the total bandwidth of the complete n-stage chain can be expressed by a combination of Formulas (7.15), (7.16) and (7.18):

$$\begin{aligned}\omega_{\text{3dB, n-stage}} &= \omega_{\text{3dB, single stage}} \cdot \sqrt{2^{\frac{1}{n}} - 1} \\ &= \frac{f_{t,\text{ input}}}{1 + \sqrt[n]{A_{\text{v, total}}}\left[k + \frac{1}{\sqrt[n]{g_{\text{mo}} R_{\text{load}}}}\right]} \cdot \sqrt{2^{\frac{1}{n}} - 1} \qquad (7.19)\end{aligned}$$

Figure 7.5 shows the bandwidth of the amplifier chain versus a sweep over the number of stages. The total required voltage gain ($A_{\text{v, total}}$) was also varied from 0 to 40 dB. In this figure it is clearly visible that two opposing effects are at work at the same time. If the number of stages is too low, capacitive load due to upscaling the amplifiers will dominate. On the other hand, the total bandwidth of the chain will drop for a higher number of stages. This despite the reduced load on each separate core amplifier. In order to get a grasp on the sensitivity of the results, it is also interesting to play around with the parameters. The main observation should be that – at least for reasonable values – the number of stages always stays within the range of $n = 3$ to 10. From this graph it can also be observed that the eight-stage amplifier is a good candidate to achieve a bandwidth of about 1 GHz in combination with a voltage gain of 30 dB. Finally, the optimum scaling factor between the stages is finally given by (7.20):

$$
\begin{aligned}
x &= 1/\sqrt[n]{g_{\text{mo}} R_{\text{load}}} \\
&= 1/\sqrt[8]{1.1\text{mS} \cdot 50\Omega} = 1.43
\end{aligned}
\qquad (7.20)
$$

Design of the core amplifier cell

The core schematic of the nonlinear loaded amplifier, this time extended with the supporting bias circuitry is shown in Figure 7.6. Instead of a single common-mode feedback circuit controlling the average voltage level at the output, both terminals have their own voltage-regulation circuit. The reason behind this choice is that DC-coupling is going to be used as the interconnect method for the subsequent core amplifiers. As was already explained in

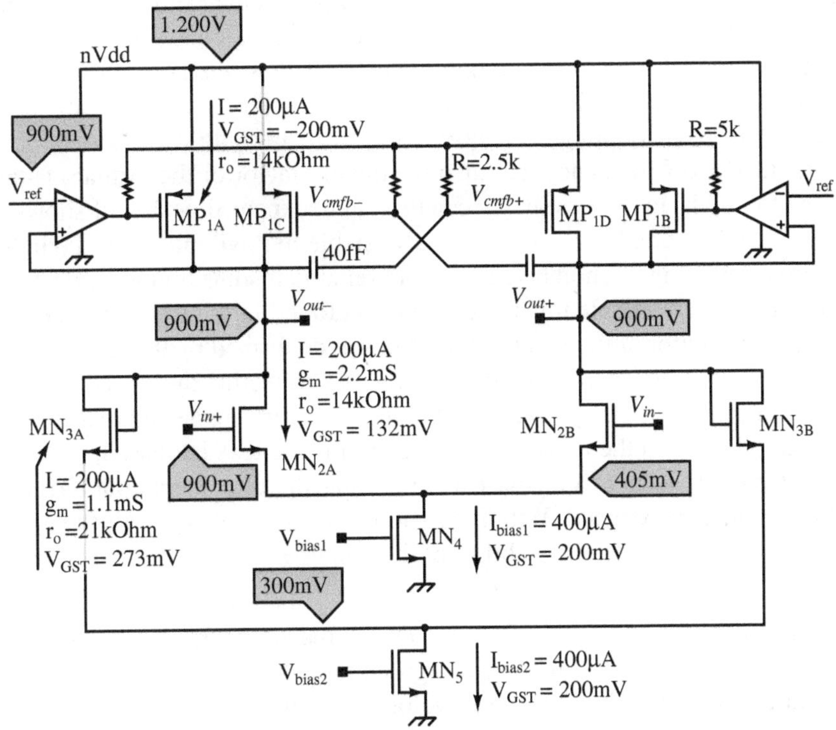

Figure 7.6. Detailed schematic of the open-loop core cell, with supporting circuitry. At the top is the differential offset compensation circuit, combined with capacitive cross-coupling to obtain some peaking around the 3dB cut-off frequency of the amplifier chain. It should be stressed that the latter measure cannot improve distortion suppression (see text).

Section 7.1, a tiny DC-offset will grow to a large problem further on in the chain. By the introduction of a separate offset-compensation circuit on each of the output nodes, deviations from the reference voltage are suppressed as long they stay within the range of the offset compensation circuit. The output of the offset compensation circuit is connected to the biasing transistors MP_{1a} and MP_{1b}.

Also note that only half of the bias current is controlled by the offset compensation circuit, while two other PMOS current sources (MP_{1c} and MP_{1d}) are driven by the averaged output level of both compensation circuits. The reason for this is nontrivial, but also quite easy to explain. Remark the two cross-coupled capacitors between the output terminals and transistors MP_{1c} and MP_{1d}. In combination with the resistors at the gates of those transistors, they form a high-frequency zero which acts as a virtual inductance connected to the output nodes of the core amplifier. This inductance can be tuned in order to obtain some form of inductive peaking near the 3 dB cut-off frequency of the amplifier.[4] Thanks to this low-quality tank, the 3 dB pole is then shifted to a slightly higher frequency. The selection of the value of the cross-coupling capacitances is quite crucial. If the cross-coupling factor is too low, the peaking will come too late so that the bandwidth of the amplifier is not extended at all. If the value of the cross-coupling capacitors is too large, peaking will not only occur before the cut-off frequency, but also the quality factor of the virtual LC-tank will be too high. In the best-case scenario, an unexpected peak will shows up in the bode diagram. In worst case, a new astable multivibrator sees daylight.[5] The cross-coupling method becomes more reliable if some tuning flexibility is added. The first available tunable capacitor is found in the form of the depletion layer capacitor of the reverse biased pn-diode formed by the drain and the n-well of transistors MP_{1c} and MP_{1d}. Instead of connecting the n-well implants of these transistors to the power supply rail, they were connected to a separate biasing node. When the reverse voltage over this diode is increased, the width of the depletion layer will increase and the capacitance at the output terminals of the amplifier is reduced. Within certain limits, the bandwidth of the amplifier can be tuned so that the stability of the amplifier can be guaranteed over a broad range of process variations.

Also shown in the schematic of Figure 7.6 are the actual transistor values that were used during the design.[6] The design journey that leads to the final layout was started by determining the bias voltages of all nodes in the circuit. First of all, the overdrive voltages of the tail current sources (MN_4 and MN_5) were assigned. The overdrive voltage should be chosen relatively high (200 mV) in

[4]This was done using repeated parasitic extractions on the layout.

[5]Provided that the chip is not packaged.

[6]Parameters of the 0.13 µm technology: $V_{supply} = 1.2$ V, $V_{Tn} \approx 260$ mV, $V_{Tp} \approx 280$ mV, $l_{min} = 0.12$ µm.

order to reduce the small-signal transconductance gain. This is done in order to obtain a low sensitivity of the bias current to variations of the gate-source voltage of the current source. For exactly the same reasons, the PMOS transistors at the top of the schematic were biased 200 mV above their threshold voltage.

The next step in the determination of the node voltages of the circuit is to divide the 1.2 V supply voltage among the transistors. The biggest chunk of the voltage headroom, 600 mV, is consumed by the diode-connected transistors (MN_3). The remaining voltage is divided between the current sources at the top of the schematic (300 mV over MP_{1a-1d}) and the tail current source (300 mV over MN_5). The common-mode level of both the input and the output of the core amplifier is 900 mV. Doing so provides enough space for a differential output swing up to 400 $mV_{diff,\ ptp}$, while none of the transistors will be pulled out of the saturation region. The only node of which the voltage remains a variable parameter is the drain voltage of MN_4, the common tail current node of the input transistor pair. This voltage automatically follows from the dimensions of the gain transistor pair. Since the ratio between the widths of the gain and loading transistors is already set by the target gain of the amplifier, the voltage at node n1 of the loading pair is not an independent parameter.

The length of the transistors was determined by their actual function in the circuit. All current sources were given a channel length of 0.65 μm. The extra parasitic drain capacitance is fairly low compared to the gate capacitor, thus by increasing their length, the output resistance is improved without too much penalty on the bandwidth of the amplifier. For the same reason, the lengths of the gain and loading pair were chosen much smaller, 0.24 μm, because their gates are connected to the input- and output-terminals and therefore a larger gate capacitor would be immediately visible in the frequency performance plot.

There is still one more final parameter has to be set before the circuit can be handed over to an automatic solver: the current through the transistors. Each of the four top PMOS current sources draws 200 μA from the power supply. The current in each branch of the differential amplifier is equally split over the gain and the loading transistor pairs. The total current of each differential pair is then combined again in the tail current source: both MN_4 and MN_5 drain in total 800 μA to the ground line of the power supply. Note that this total current was determined based on a target output impedance ($1/g_{mo}$) of around 900 Ω of transistors $MN_{3a,b}$.

After running the automatic optimizer, the target widths of the transistors become available and the process of fine tuning can be started. First of all, the voltage level on node n1 of the loading pair must be verified, since this was the only node of which the voltage was not explicitly set. As can be seen in the schematic (Figure 7.6), the voltage level on this node is 405 mV, which drives the drain voltage of MN_4 about 200 mV in the saturation region. More important is the overdrive voltage of the input transistor pair, as it will limit the signal swing at the input of the amplifier. Since the overdrive voltage of the input pair is 132 mV, a differential input swing of more than 400 mV_{ptp} can be properly handled. Also note that the output resistance of each transistor that is connected to the output terminals of the amplifier was kept above 10 kΩ. This way, the total value of unwanted parasitic output resistance makes up less than 20% of the impedance of the nonlinear load ($1/g_{mo}$). As a final remark, it should be taken into account that in the tapered buffer setup (scaling factor $x = 1.45$), the total capacitive load will increase by around 10%.

Layout of the tapered buffer

During the entire design process, economy of time and effort was kept in mind. The layout was started by the construction of a high-level floorplan where the number and position of the bonding pads was determined. The amplifier has in total four high-frequency in- and outputs. At the off-chip level, interfacing with each of these lines occurs with (large) microstrip lines which transport the signal as close as possible to the bonding pads of the chip. For this reason, each of the four IO-lines was given a central position at one of the edges of the chip die (Figure 7.8). The remaining DC-carrying nodes are located alongside the central terminals: under the ideal circumstances, not only the actual ground terminal, but also the power supply and biasing lines can act as a virtual ground shields for the AC-input signals.

The layout of the chip is based on a modular design, using only four basic building blocks: an ESD subcell, a power supply decoupling cell, an offset compensation subcircuit and finally also a minimum sized core amplifier. The layout of the core amplifier blocks is arranged in such a manner that two core cells can be joined together from any direction. The layout structure of a single core cell is shown in Figure 7.7. Also shown in this layout are all external connections from this cell to the outside world. Remark that the same interconnect lines appear at each of the four sides of the amplifier. Doing this way, it becomes possible to stack several amplifier modules together in order to form a new amplifier with larger driving capabilities.

Instead of starting a new design with the appropriate dimensions for every of the eight stages in the amplifier chain, the required number of core amplifier cells is sticked together at the next higher abstraction level in the layout. No extra interconnect effort is needed thanks to the modular approach of the core cells. The first stage in the chain embodies only two amplifier cells. Based on the upscaling factor of $x = 1.45$, the next stage contains three cells and so on until the last stage, which contains a 16-valve core engine. The offset

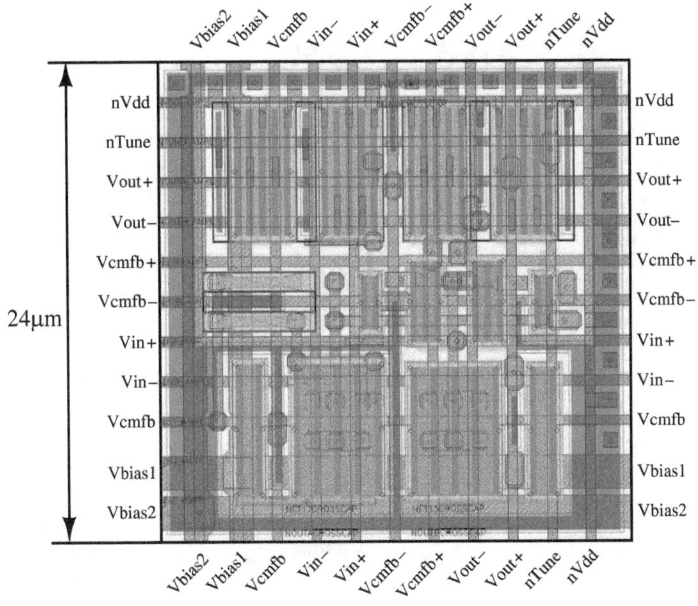

Figure 7.7. Layout of the open-loop core cell module of Figure 7.6. Several of these core cells can be stacked to each other to form an amplifier stage with higher driving capabilities. A total of 58 modules were used in the eight-stage amplifier.

compensation circuit is based on the same principle: it can be plugged to any of the edges of a core amplifier assembly. Most of the area of the offset compensation circuitry is taken into account by the low-pass filter, which defines the 3 dB corner frequency at the lower end of the spectrum. Signal frequencies below this pole will be treated as offset and are thus suppressed.

At the next level in the design hierarchy, the remaining space in the active area of the chip is filled up with decoupling modules which are divided among the power supply and biasing lines of the amplifier chain. The total area of all stacked amplifiers, offset regulators and decoupling capacitors adds up to $0.38 \times 0.38\ \text{mm}^2$. Only one hierarchical level in the layout has to be completed by now: the ESD protection layer is placed in between the baseband amplifier in the center of the chip and the bonding pads. Fully compatible with the rest of the design, the protection ring is also built around a basic platform of stackable ESD modules. With only minimal intervention, the base ESD unit can be adapted to fit the needs of a power supply input, a biasing terminal, or an IO-line. For example, if the ESD structure is connected to one of the high-frequency input lines of amplifier, a 50 Ω termination resistor can be placed in a special notch of the base ESD module.

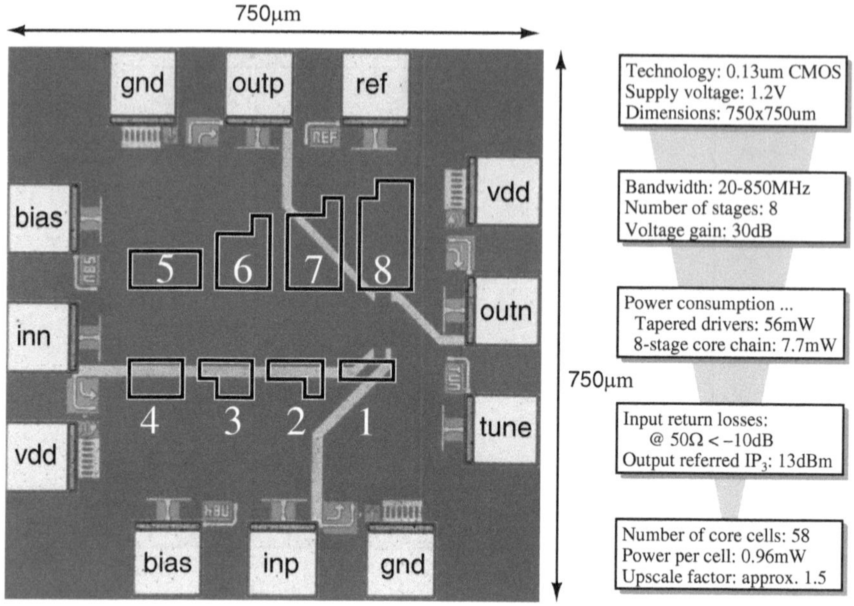

Figure 7.8. Chip microphotograph of the eight-stage open loop amplifier, with a layout overlay showing the tapered buffer structure. Each subsequent stage in the buffer is upscaled with a factor of about 1.5, by simply parallelizing multiple core amplifier cells from Figure 7.6. The first stage embeds two core cells, while the last stage contains as much as 16 identical cells.

The final layout of the chip is shown in Figure 7.8. It shows the microphotograph of the chip with a layout overlay which indicates the tapered eight-stage buffer structure.[7]

Measurement setup and experimental results

The proposed baseband amplifier was fabricated in a 0.13 μm standard CMOS process. The chip is directly mounted on an alumina substrate (Figure 7.9) with embedded grounded microstrip structures. The external components are limited to a single supply decoupling capacitor and two balun transformers to transform the differential in- and outputs to the single-ended measurement setup. The active section of the chip includes 58 core cell amplifiers and draws in total 56 mW from the 1.2 V power supply. This is because of the tapered setup of the eight-stage amplifier, necessary to drive the measurement equipment. Of course, when the amplifier is used in an embedded system, this

[7] Which is not visible due to the ground and power planes which cover most of the upper metal layers.

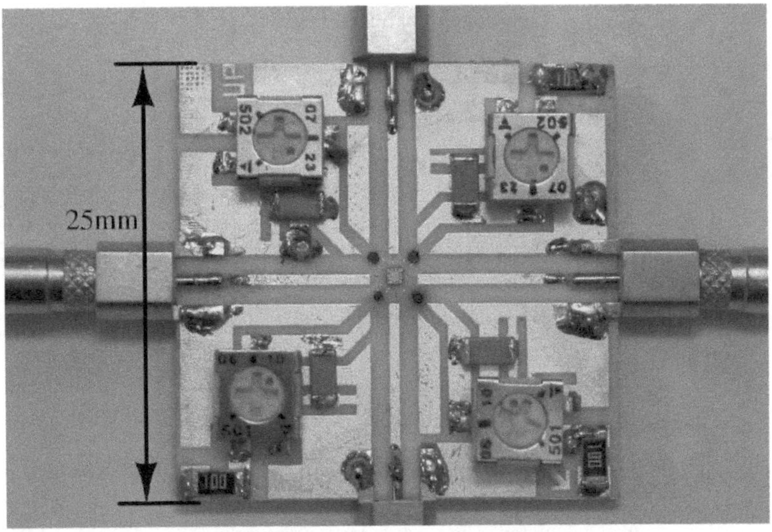

Figure 7.9. Photograph of the measurement setup. The chip is bonded on an alumina substrate. Note the four microstrip structures which deliver the RF-signals as close as possible to the amplifier. The ground plane of the microstrip is not visible because it is at the back of the substrate.

removes the need of using a tapered buffer and as such saves chip area and power. Each stage provides a voltage gain of 1.5 (3.5 dBV), resulting in an overall gain of about 30 dB. The measured 3 dB bandwidth of the amplifier ranges from 20 up to 850 MHz, which could be achieved thanks to the capacitive cross-coupling near the −3 dB frequency.

The measured bandwidth performance of the amplifier is shown in Figure 7.10. During the first measurements, reflections at the input stage of the amplifier caused a considerable amount of ripple in the frequency response at the higher end of the frequency band. It turned out that the problem was caused by the bonding wires which interconnect to the microstrip structures in combination with the capacitive load of on-chip ESD protection structures. This assumption was also confirmed by S_{11} measurements at the input of the amplifier. Because of the large fractional bandwidth of the system, it becomes very difficult to tune this inductance to the frequency of interest with traditional on-chip matching methods. The issue was solved by using shorter bonding wires and by laser-cutting part of the ESD protection from the input stages. An even better solution would consist in a complete switch-over to a flip-chip approach.

For a gain setting of 30 dB, the output-referred third-order intercept point (OIP_3) over the specified frequency band is better than +11 dBm measured over a 50 Ω load impedance. A summary of the measurement results can be

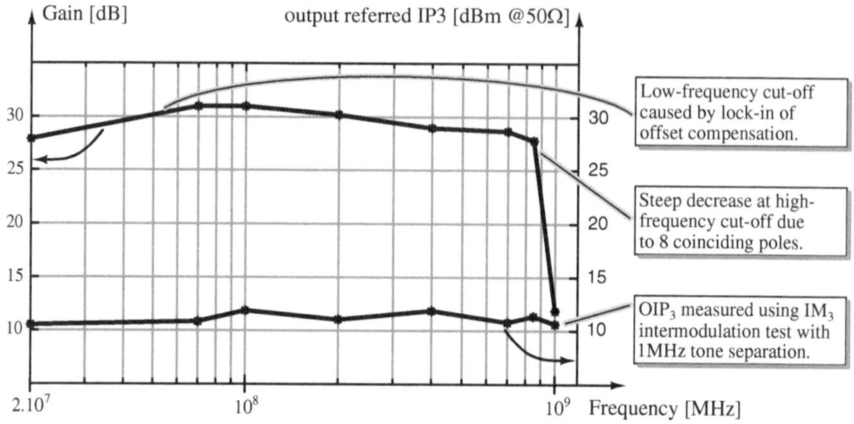

Figure 7.10. Measurement result of the eight-stage open-loop amplifier. The output-referred third-order intermodulation intercept point (OIP_3) was calculated using a two-tone intermodulation test with 1 MHz separated frequencies. The OIP_3 was the obtained using $OIP_3 = P_{out} - \frac{1}{2}IM_3$.

found in Figure 7.8. The proposed nonlinear loaded open-loop amplifier outperforms comparable structures reported in the literature [Lee06, Duo06].

7.7 Overview and future work

The nonlinear characteristics of an active element can be improved by embedding this element in the forward path of a feedback system: a large loop gain makes that the transfer function of a closed-loop system is determined by the characteristic of the feedback circuitry,[8] rather than the non-ideal transfer function of the active element. What is often overlooked, though, is the fact that distortion suppression only starts when some excess loop gain becomes available to the system (Appendix A.2). Only starting from below the closed-loop cut-off frequency, distortion is increasingly suppressed, down to the first pole of the open-loop system.

This finding may lead to different conclusions, depending on the specific application conditions of the feedback amplifier. If the amplifier is going to be used in a low-speed high-accuracy circuit such as a sensor or an audio preamplifier, the low-frequency gain of the amplifier must be as high as possible, while the 3 dB cut-off frequency of the *open-loop* system must be of the same order of magnitude as the bandwidth that is required by the application. In

[8] Because of their decent linearity performance, passive devices are commonly used in the feedback path.

analog CMOS circuit design, this goal is typically achieved using two-stage amplifier, of which the low-frequency gain is even more increased using cascoded or gain-boosted voltage amplification stages. Beyond this frequency, the distortion performance drops off steeply, as is pointed out in Appendix A.2.

Unfortunately, the bandwidth of the feedback-oriented approach is somewhat limited. This is because a high voltage gain at a certain circuit node is always accompanied by high node impedances, which in turn leads to a lower cut-off frequency of the open-loop system. For medium to high-speed applications, which operate in the frequency range between the first pole of the active element and the first pole of the closed-loop system, the simultaneous increase in the DC-gain and reduction of the open-loop bandwidth will not lead to an improved distortion suppression. The reason for this is that the performance of such systems is commonly expressed using the *effective resolution bandwidth* (ERBW), which takes into account that the amplifier must achieve the required linearity performance, even at the maximum frequency-of-interest. However, exactly the lack of excess gain at higher frequencies make the feedback amplifier unsuited for application in, for example, the VGA of the pulse-based radio receiver in Chapter 6.

This chapter tries an unconventional approach, by compensating the nonlinear transconductance of the CMOS gain pair with a nonlinear load. In this amplifier, the current-to-voltage conversion step at the output is performed by an active load, which has the same transfer characteristics as the gain element. It was shown that, for certain settings of the operating point of the gain- and the load-transistors, the third-order harmonic distortion of the nonlinear loaded amplifier can be completely cancelled by this approach. Since the exact settings for optimal distortion performance depend on several parameters, such as the parasitic output resistance and the matching parameters of the underlying CMOS technology, calibration is a crucial factor for the performance of the proposed amplifier.

Suggestions for improvements

In the current implementation of the standalone open-loop amplifier, the chip directly drives the 50 Ω load impedance of the measurement setup. This in contrast to an amplifier that is embedded in a system and only drives the capacitive load of an on-chip AD converter. In the former case, however, the linear current-voltage characteristic of the linear output impedance leads to the regeneration of distortion components in the very last stage of the multistage amplifier. This may mask the true capabilities of the open-loop approach. A much better approach is to measure the performance of the amplifier in an embedded application, where the wideband buffer stage at the output is replaced by a downconversion mixer and a very linear narrowband feedback-based output buffer stage. Instead of a direct measurement of the entire frequency range, the band-of-interest is first downconverted to the passband frequency of the buffer stage. A sweep over the frequency band provides a reliable indirect measurement of the linearity of the open-loop amplifier.

Also, the chip described in Section 7.6 does not provide a satisfactory means to tune its gain during operation. Since the amplifier was designed to achieve optimal distortion performance for a gain setting of 30 dB, changing the ratio between the tail currents of the gain- and the loading pair would move the bias point away from the optimal settings for distortion suppression. The correct way to introduce a tunable gain in the system is to set the last few stages to a fixed gain and implement a limited amount of gain tuning in the first few stages. The validity of this suggestion is verified by recognizing that the final stages of the amplifier are most susceptible to a deviation on their tail current ratio, as they must handle the largest signal swing. On the other hand, the first few stages have only a small signal swing over their output terminals. As a result, a limited offset from the optimal bias point can be tolerated without adverse consequences for the overall performance of the amplifier. One must keep in mind that reducing the gain of the very first stage should be avoided, since this may affect the noise figure of the amplifier. The contribution of the successive stages to the noise floor, however, is reduced by the voltage gain of the first stage.[9]

The picture at the right shows a close-up of the amplifier which is wire-bonded on an alumina substrate. One of the problems encountered during the measurements was the parasitic impedance of the bonding wires that connect the microstrip lines with the input bonding pads.[10] In combination with the capacitance of the on-chip ESD structures, the inductance of the bonding wires (1–2nH) causes reflections which is seen as ripple at the higher end of the frequency spectrum (from 750 MHz and up). In the measurement results of Section 7.9, this effect was eliminated by taking the S_{11} reflection coefficient into account. In a practical application, the effect of the bonding wires should be moved out of the signal band. The wide fractional bandwidth of the system, however, prevents that the parasitic bonding wire inductors can be tuned to the frequency band of interest. A possible solution might be to eliminate bonding wires and turn to a flip-chip approach. Nonetheless, the flip-chip technique is not for the faint-of-heart, so the author left it for future generations to investigate.

[9]Further information on the noise figure of a cascaded system can be found in Section 6.3.

[10]The 50 Ω termination resistors are located on-chip.

The current level of implementation requires a certain degree of calibration to realize the best results. The end user, however, does not want to be confronted with this precarious task. For this, the barebone open-loop amplifier setup can be extended with an on-chip monitoring circuit to guarantee good performance over a large range of circumstances (temperature and process variations, varying power supply levels). Fortunately, if the amplifier is going to be embedded in the signal path of a digital receiver, most of the complexity required for this purpose is already on-board, in the form of a digital signal processor. For a fully-featured automated calibration system, the setup needs to be extended with an on-chip two-tone generator. The only thing left to do for the signal processor in the back-end is to monitor some interesting frequencies such as the third-order intermodulation frequency of the two test tones. Based on measurements of the spurious signal levels at this frequency, the ratio between the tail currents of the gain- and loading pair can be fine-tuned in order to obtain the best possible biasing setting. The calibration itself could be done while the receiver is online, by using some unused frequency bands (for example below the 3 dB cut-off frequency at the lower end of the spectrum) or off-line, during idle periods in between data reception. In the meanwhile, though, the manual calibration method was just enough to demonstrate the promising capabilities of the prototype open-loop amplifier.

Appendix A
Distortion analysis of feedback amplifiers

In this appendix, the theory of distortion suppression in feedback systems is reviewed briefly. Feedback is commonly used to suppress the non-ideal characteristics of the active gain stage, such as a poorly defined gain factor or a nonlinear transfer function between in- and output. The ability of feedback to mask the imperfections of the gain stage rely on the availability of a certain amount of excess loop gain. Unfortunately, exactly the lack of loop gain at higher frequency bands prevents feedback-based amplifiers to be used in wide-band applications. In order to get some insight in this matter, this chapter starts with a general review of the theory on feedback and linearity. While the calculations in the first few sections will deliberately ignore all non-ideal effects which are encountered in practical amplifier implementations, non-idealities such as a finite output resistance of the transistors, bandwidth limitations or the issues caused by stability requirements are gradually introduced and discussed in more detail in the subsequent sections.

To get off to a good start, consider a transistor amplifier that is used within it's weak nonlinear operating region. The time-domain input signal is represented by $x(t)$ and the output signal of the amplifier is given by $y(t)$. The frequency independent relationship between the in- and output of this system is being described by (A.1):

$$y(t) = a_0 + a_1 x(t) + a_2 x^2(t) + a_3 x^3(t) + \dots \tag{A.1}$$

Coefficient a_0 represents the DC-current at the output of the amplifier, which is independent from the signal applied to the input. At first sight, the DC-output of the amplifier may seem a somewhat pointless detail, but it will become clear later on that this offset voltage is a major headache for the designer of an open-loop based system. For a single-ended transistor stage, the DC-offset is determined by the operating point of the transistor. In the majority of the cases, the output voltage of an amplifier is not compatible with the operating point of the subsequent stage. Because the in- and output operating point voltage levels are commonly fixed at different voltages, a DC-coupling between subsequent stages of a multistage amplifier becomes highly problematic. Then again, in the small number of cases where the stages are directly coupled, one must proceed with extreme caution, since a small deviation from the operating point is treated as a valid input signal by the next stage. This causes the amplifier to drift away from the intended biasing point, which may cause clipping in the next stage. It is for this reason that the DC-voltage is removed from the signal path in almost every high-gain amplifier (e.g. the IF-stages of a receiver). This is commonly achieved by employing AC-coupled circuit sections: coupling capacitors block the DC-component and pass the higher frequency components to the next stage.

Coefficient a_1 from Equation (A.1) represents the linear gain of the amplifier. For example, the large-signal model for a MOS transistor in strong inversion is described by (A.2):

$$I_{DS} = K'\frac{W}{L}(V_{GS} - V_T)^2 , \tag{A.2}$$

where I_{DS} is the drain-source current as a result of the V_{GS} voltage applied over the gate-source terminals of the transistor. The W/L ratio is defined by the dimensions of the transistor and V_T is the threshold voltage. The transconductance parameter, K' (units: A/V^2), depends on technology parameters such as electron mobility and oxide capacitance. These characteristics are generally obtained from measurements. The small-signal linear gain of the MOS amplifier can be obtained by choosing the DC-operating point of the MOS transistor in the strong inversion region and then applying a sufficiently small voltage swing around this operating point. The first order variation of the drain current is then given by:

$$a_1 = g_m = \frac{i_{DS}}{v_{GS}} = \frac{2 \cdot I_{DS}}{V_{GS} - V_T} , \tag{A.3}$$

where v_{GS} is the small-signal input voltage and i_{DS} is the resulting output current. The fraction i_{DS}/v_{GS} is called the transconductance g_m (unit: A/V) of a MOS transistor. The capitalized parameters V_{GS} and I_{DS} define the large signal operating point. One can see that the first order transconductance can be increased in two ways. The first option is an increase of the drain-source current flowing through the transistor. If the overdrive voltage ($V_{GS} - V_T$) needs to remain fixed, this is achieved by changing the W/L dimensions of the transistor accordingly. On the other hand, if the bias current needs to be kept fixed, a reduction of the overdrive voltage across the gate-source terminals results in an inverse proportional increase of the transconductance. Remark that the dimensions of the transistor should be scaled appropriately to maintain a fixed bias current.

At first glance, one would prefer a reduction of the overdrive voltage, while the bias current remains unaltered. After all, this method seems to increase the DC-gain of the transistor without raising the power consumption. Unfortunately, there is no such thing as a free lunch. In order to clarify this universal misfortune, let us first introduce some important performance parameters of the MOS transistor.

CUT-OFF FREQUENCY OF A MOS TRANSISTOR

The *cut-off frequency* (f_T) of a single MOS transistor is defined as the frequency at which the small-signal gate current becomes equal to the small-signal drain current (with drain shorted to AC-ground). It is the transition frequency (hence the abbreviation f_T) beyond which the transistor does not provide any useful current gain. The cut-off frequency is a measure for the intrinsic speed (excluding junction and wiring impedance) of a transistor and is given by (A.4):

$$f_T = \frac{g_m}{2\pi C_{GS}} \text{ [Hz]}, \tag{A.4}$$

where C_{GS} is a rough estimation of the parasitic gate-source capacitance. In the strong inversion region, this capacitance is dominated by the oxide capacitance between the gate and the inversion layer at the source node implant of the MOS transistor [Eyn89, Raz94].

MAXIMUM FREQUENCY OF OSCILLATION

The *maximum frequency of oscillation* (f_{max}) of a MOS device is defined as the maximum frequency at which the unilateral power gain[1] of a transistor drops below 0 dB [Dro55]. The f_{max} limit can be reached if the impedance of parasitic capacitances at gate and drain are compensated using appropriately sized inductors.[2]

[1] Unilateral power gain: the power gain of a transistor when feedback has been used to neutralize S_{12}. Also, both the in- and the output reflection coefficients (S_{11} and S_{22}) have been matched to zero.

[2] Such as is done in an oscillator, hence the term *maximum frequency of oscillation*.

If the source impedance is chosen equal to the gate resistance r_G and the load impedance is perfectly matched to the small-signal output resistance r_{DS}, the power gain G of this circuit can be calculated as (A.5):

$$G = \frac{g_m^2 \, r_{DS}}{4\,\omega^2 \, r_G \, C_{GS}^2}, \tag{A.5}$$

which can also be expressed using the cut-off frequency f_T as (A.6):

$$G = \frac{r_{DS}}{4\,r_G} \cdot \left(\frac{f_T}{f}\right)^2 \tag{A.6}$$

The maximum frequency of oscillation is reached when the power gain drops below 0 dB:

$$f_{max} = \frac{f_T}{2} \cdot \sqrt{\frac{r_{DS}}{r_G}} \; [\text{Hz}] \tag{A.7}$$

Since a MOS transistor operated in the saturation region is a transconductance-mode device, this result is not surprising: the series gate resistance r_g, in combination with the gate capacitance C_{GS}, produces a high frequency pole.[3] Beyond this frequency, the effective voltage that is available over the gate-oxide capacitor C_{GS} starts to decrease, thereby setting a limit on the power gain-bandwidth product (f_{max}). It follows that, for high-frequency performance, the gate resistance should be as low as possible. For this reason, a low-resistance salicided polysilicon layer[4] is formed above the gate of a MOS transistor.

Because a MOS transistor in saturation is a voltage controlled current source, more power is delivered to a higher load-impedance. This is also found in Formula (A.7), since the maximum frequency of oscillation increases for a higher load resistance r_{DS}. It also follows that, depending on the r_{DS}/r_G ratio, f_{max} may be higher or lower than the cut-off frequency of the transistor. This is quite an interesting observation because it raises the question which parameter fits best for a particular circuit setup.

The circuit of a basic multistage transconductance amplifier is shown in figure A.1. Each of the identical stages is build around an NMOS common source

[3] Remark that, although it is omitted here, the resistance of the inversion layer itself also plays an important role in the effective gate resistance.

[4] Basically, this is a metal-silicide contact.

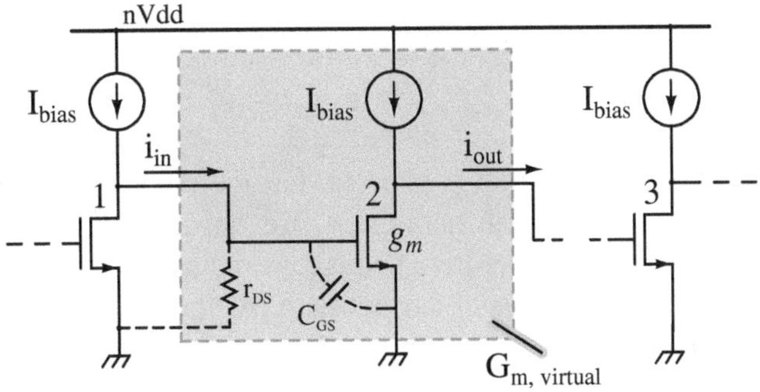

Figure A.1. A basic multistage transconductance amplifier. The gray-shaded area represents a virtual current-gain amplifier which includes the parasitic output resistance of the previous stage and the gate capacitance of the current stage. For reasons of simplicity, the DC-operating point of the transistors is ignored.

amplifier. Suppose that the transistors are biased in their saturation region, and the output impedance of the ideal current source is sufficiently high so that it be neglected.

For low frequencies, the small-signal output current i_{out} of each stage is absorbed by it's own output resistance r_{DS}. At the same time, the voltage drop over this resistor is also the small-signal input voltage of the following stage. The current gain of each virtual intermediate stage (highlighted by the shaded area in Figure A.1) is thus given by $g_m r_{\text{DS}}$. For increasing frequencies, an increasing portion of the input current is dumped in the parasitic gate-source capacitor C_{GS}, until the cut-off frequency f_T has been reached. Apart from consuming power, the intermediate stage does not provide any additional benefit beyond this frequency point.

Back to the original question posed in the beginning of this section: should the transconductance g_m be increased by reducing the overdrive voltage or by increasing the bias current? As an experiment consider the case where $V_{\text{GS}} - V_T$ is reduced. If the bias current I_{bias} of the transistors remains constant, if follows from Equation (A.3) that g_m is inverse proportional to the overdrive voltage:

$$g_m \propto \frac{1}{V_{\text{GS}} - V_T}, \quad \text{for fixed } I_{\text{bias}} \qquad \text{(A.8)}$$
$$\text{thus: } g_m \nearrow \text{ if } V_{\text{GS}} - V_T \searrow$$

On the other hand, it follows from Equation (A.2) that the width of the transistor needs to be increased in order to maintain a fixed bias current. As a result, the gate-source capacitance C_{GS} of the transistor is increased by the same ratio:

$$C_{\mathrm{GS}} \propto W \propto \frac{1}{(V_{\mathrm{GS}} - V_T)^2}, \text{ fixed } I_{\mathrm{bias}} \tag{A.9}$$
$$\text{thus: } C_{\mathrm{GS}} \nearrow\!\!\nearrow \text{ if } V_{\mathrm{GS}} - V_T \searrow$$

The answer to the first question is twofold. For the circuit depicted in Figure A.1, the low frequency current gain ($g_m r_{\mathrm{DS}}$) of an intermediate stage is inversely proportional to the overdrive voltage. A reduced overdrive voltage results in an improved small-signal current gain. At the high frequency side of the spectrum, the opposite option is more advantageous. From the definition of f_T and from Equations (A.8), (A.9), it follows that the current gain-bandwidth product f_{T} increases proportionally with an increased overdrive voltage V_{GST} (A.10):

$$f_T = \frac{g_m}{2\pi C_{\mathrm{GS}}} \propto V_{\mathrm{GS}} - V_{\mathrm{T}}, \text{ for fixed } I_{\mathrm{bias}} \tag{A.10}$$

The reader should keep in mind that the simplified strong inversion model of A.2 is not entirely accurate in a deep-submicron CMOS technology. The I_{DS} versus V_{GS} characteristic becomes less quadratic for smaller transistor lengths. Ultimately, beyond the velocity saturation point where the characteristic becomes linear, both g_m and f_T both become independent of the overdrive voltage. The general idea of the above findings, however, remains valid for the complete spectrum of models in between the quadratic and a more linear transistor model.

The second experiment involves the case where the overdrive voltage $V_{\mathrm{GS}} - V_T$ stays constant, but this time only the DC bias current I_{bias} through the transistor is increased. Once again using Equation (A.3), it follows that g_m is proportional to the bias current. If the bias current through each section of the multistage amplifier is scaled evenly, it is interesting to notice that the low frequency current gain $g_m r_{\mathrm{DS}}$ of the intermediate stage remains more or less constant: the width W of the transistor scales proportional to the bias current and results in an inverse proportional decrease of the output resistance of each stage:

$$\begin{aligned} g_m &\propto I_{\mathrm{bias}} \\ r_{\mathrm{DS}} &\propto \frac{1}{W} \propto \frac{1}{I_{\mathrm{bias}}} \\ g_m r_{\mathrm{DS}} &\approx \text{constant}, \text{ for fixed } V_{\mathrm{GS}} - V_T \end{aligned} \tag{A.11}$$

The same conclusion can be made for the cut-off frequency of an intermediate transistor stage. Both the transconductance and the parasitic oxide capacitance are in first-order approximation proportional to the bias current, which results in a bias current independent gain-bandwidth product f_T:

$$
\begin{aligned}
g_m &\propto I_{\text{bias}} \\
C_{\text{GS}} &\propto W \propto I_{\text{bias}} \\
f_T &\approx \text{constant, for fixed } V_{\text{GS}} - V_T
\end{aligned}
\tag{A.12}
$$

The results presented here should not be surprising after all, since scaling a transistor's width W without changing the length L basically boils down to placing several independent transistors in parallel. Intuitively one can see that connecting those transistors in parallel should have no effect on gain or on frequency performance. To conclude this introduction, Table A.1 summarizes the considerations one should make during the design of an analog MOST amplifier. Although the deductions were made using the simplified model of a MOS transistor in the strong inversion saturation region, the general idea remains valid for characteristics anywhere between a quadratic and a linear current-voltage transistor model.

	DC-gain ($g_m r_{\text{DS}}$)	Cut-off frequency (f_T)
Fixed I_{bias}, variable V_{GST}	$\propto \frac{1}{V_{\text{GS}}-V_T}$	$\propto V_{\text{GS}} - V_T$
Fixed V_{GST}, variable I_{bias}	Constant	Constant

Table A.1. Correlation between biasing and small-signal performance of a MOS transistor in the strong inversion saturation region.

Remark that for a fixed overdrive voltage $V_{\text{GS}} - V_T$, both the DC and the high frequency performance of the MOS transistor are independent of the power consumption. Only technological parameters such as oxide thickness (t_{ox}), dielectric constant (ϵ_{ox}) and minimal gate length (L_{min}) define the performance limits of a certain CMOS process.

A.1 Feedback amplifiers

Despite the good low frequency gain performance, the multistage current amplifier introduced in Figure A.1 is not suited for high frequency applications. This is because the location of the 3 dB cut-off frequency of a single stage is defined by r_{DS} and C_{GS}. The overall transfer function of a n identical cascaded gain stages is given by (A.13):

$$
A(j\omega) = \left(\frac{g_m r_{\text{DS}}}{1 + j\omega r_{\text{DS}} C_{\text{GS}}} \right)^n \tag{A.13}
$$

It follows that, for a system with n (>1) multiple coincident poles, the 3 dB bandwidth ($\omega_{3\text{dB}}$) does not correspond with the frequency pole of a single

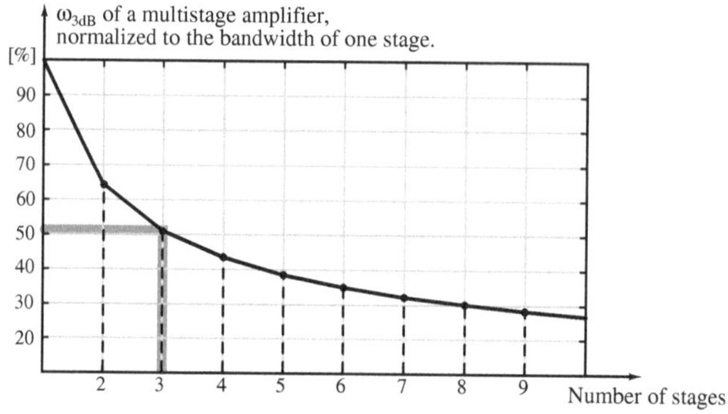

Figure A.2. Bandwidth reduction in a multistage amplifier. For example, in a cascaded amplifier with three equal gain elements, the cut-off frequency drops to 50% of that of a single stage amplifier.

stage. As shown in Figure A.2, the bandwidth of a cascaded amplifier is seriously degraded for an increasing number of gain stages (A.14):

$$\omega_{3\text{dB, n-stage}} = \frac{1}{r_{DS}C_{GS}} \cdot \sqrt{2^{\frac{1}{n}} - 1} \tag{A.14}$$

For example, suppose a deep submicron technology (e.g. 130 nm) with an f_T of 100 GHz. The low frequency gain of a single transistor is 30 dB. The bandwidth of one single ideal stage (Figure A.1) is thus 3 GHz. For a three-stage amplifier however, the bandwidth has already decreased to 1.5 GHz, even without taking the wiring or load capacitance at the output into account. In addition, both the current gain and the location of the 3 dB frequency point are defined by r_{DS} and C_{GS}. The value of these transistor parameters strongly depends on the process characteristics. It is advised against using them as a reliable or critical design constant.

The frequency performance of the multistage amplifier can be increased by applying a feedback path which spans over one or more gain stages: assuming that stability is correctly taken into account, gain can be traded for bandwidth within certain boundaries. Consider the feedback amplifier in Figure A.3. The transfer characteristic can be calculated as (A.15):

$$A_{CL} = \frac{A}{1 + AH}, \tag{A.15}$$

where A is the forward gain of the amplifier and H is the feedback factor, of which the power gain $|H|^2$ is usually smaller than unity. The transfer function

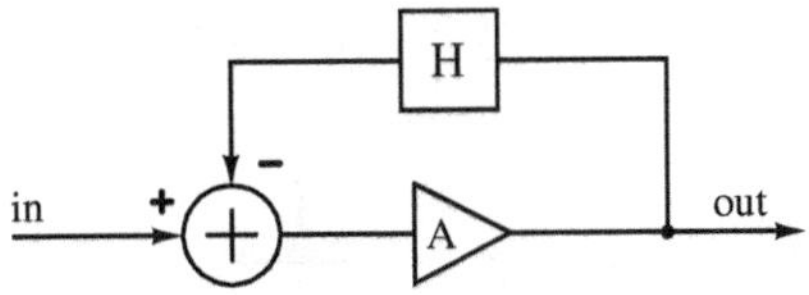

Figure A.3. Basic feedback amplifier. Factor A is the active gain element of the feedback circuit, while H is called the feedback element. For linearity reasons, the feedback element is a passive circuit in most applications.

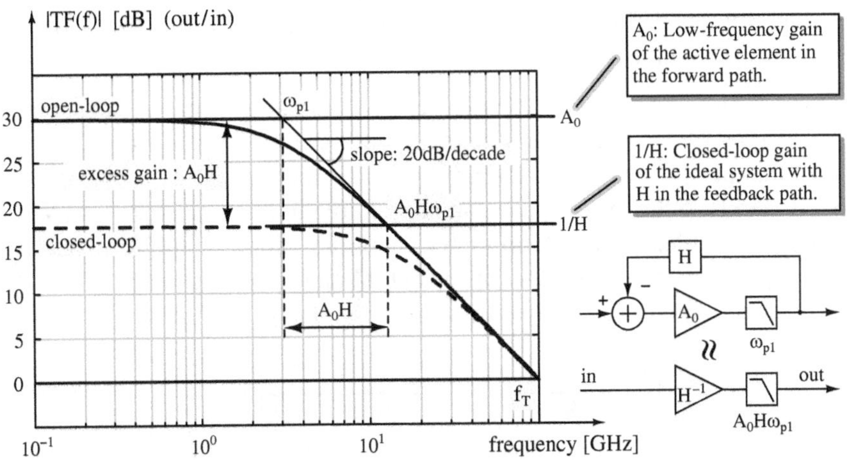

Figure A.4. Transfer characteristic of an amplifier with a single pole in an open loop (solid) and in a closed-loop (dashed) configuration. In closed-loop configuration, the excess gain suppresses non-ideal characteristics of the active element in the forward path.

of this feedback system can be approximated by $1/H$ for sufficiently large values of the forward gain factor A. As mentioned before, a practical amplifier implementation exhibits one or more poles beyond which the gain factor A starts to decrease. An amplifier setup with a single pole filter in the forward path is introduced in Figure A.4. This time, the transfer function of the single pole feedback system is given by Equation (A.15), where the ideal amplifier A is substituted by it's frequency dependent counterpart:

$$A(f) = \frac{A_0}{1 + j\frac{\omega}{\omega_{p1}}} \tag{A.16}$$

For a sufficiently large DC loop gain ($A_0 H \gg 1$), the application of a single-pole (ω_{p1}) amplifier in a feedback system results in a new single-pole system, but this time the cut-off frequency of the compound system is given by $\omega_{\text{cut-off}} = A_0 H \omega_{p1}$ (A.17):

$$A_{\text{CL}}(f) = \frac{1}{H} \cdot \frac{1}{1 + j\omega \frac{1}{A_0 H \omega_{p1}}} \tag{A.17}$$

A summary of the transfer characteristics of both the open-loop amplifier and the same amplifier in feedback configuration is given in the magnitude plot of Figure A.4. When applied to the single transistor current amplifier, the 0 dB crossover point is defined by the f_T of the MOS transistor, while the open-loop gain A_0 is then defined by $g_m r_{\text{DS}}$.

The excess gain factor $A_0 H$, shown in Figure A.4, can be employed to suppress imperfections of the active gain element. In a closed loop topology, nonlinearities that are present in the characteristics of the transistors are being suppressed thanks to the excess gain that is available in the feedback loop. While the gain of the ideal closed loop system is only defined by the feedback factor $(1/H)$, the relative deviation of the actual closed loop gain A_{CL} from this $1/H$-characteristic is determined by the excess loop gain (A.18):

$$\begin{aligned} \epsilon_{\text{rel}} = \left| \frac{1/H - A_{\text{CL}}}{1/H} \right| &= \left| \frac{1}{1 + A(j\omega)H} \right| \\ &= \left| \frac{1}{1 + A_0 H} \cdot \frac{1 + j\frac{\omega}{\omega_{p1}}}{1 + j\frac{\omega}{A_0 H \omega_{p1}}} \right| \end{aligned} \tag{A.18}$$

The frequency response of the error curve is shown in Figure A.5. At the lower end of the frequency spectrum, the relative error on the gain is solely defined by $1/(1+A_0 H)$. However, at the first pole ω_{p1} of the active gain stage in the forward path of the amplifier, the deviation from the correct gain factor has already risen to 1.4 times the error at DC. For even higher frequencies, the relative error gradually rises to 100%, which in fact means that there is no output signal at all.

Distortion in feedback amplifiers

Not only the unpredictable and finite gain A_0 of the non-ideal amplifier is suppressed in a closed-loop system, but also the linearity requirements of the amplifier are relaxed. For an increasing open-loop excess gain $A_0 H$, the nonlinear characteristics of the closed-loop system are left to the – commonly passive – elements in the feedback path. It should be easy to understand that the extent

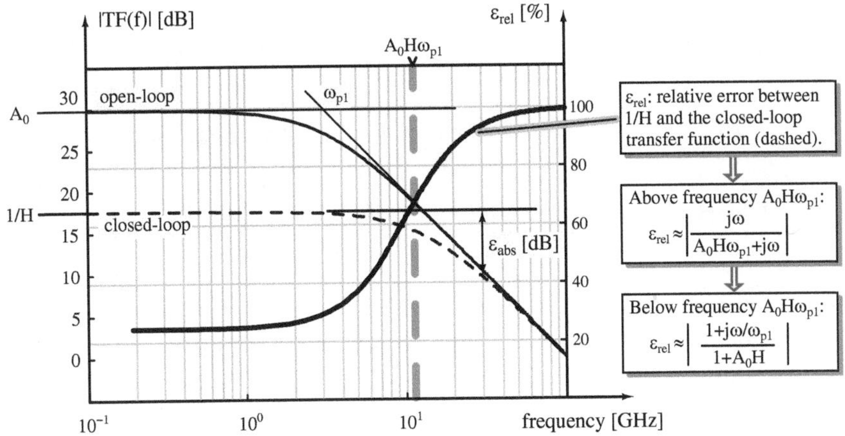

Figure A.5. Gain error between the ideal closed-loop transfer function ($1/H$) and the behaviour of a system with limited gain- and bandwidth-resources (dashed). The gain error crosses the 50% boundary before the 3 dB frequency of the closed-loop system.

in which the nonlinearities are suppressed is dependent on the frequency and, as a consequence, are related in some degree to the excess gain of the loop. As a place to start, consider a generic nonlinear *open-loop* amplifier, of which the gain factor exhibits both second- and third-order harmonic distortion. The frequency independent characteristic of the amplifier can be modelled by the following truncated polynomial (A.19):

$$a(x) = a_0 + a_1 x + a_2 x^2 + a_3 x^3 + \dots, \tag{A.19}$$

where a_0 is the DC-offset, a_1 represents the linear gain parameter and coefficients a_2 and a_3 generate the unwanted second- and third-harmonic distortion components in the output spectrum of the amplifier. When a sinusoidal wave $x = U\cos(\omega t)$ is applied to the input, the output signal of the amplifier in Formula (A.19) is given by (A.20):

$$\begin{aligned} a(x) = \left(a_0 + \frac{a_2}{2}U^2\right) \;&+\; \left(a_1 + \frac{3}{4}a_3 U^2\right) U\cos(\omega t) \\ &+\; \frac{a_2}{2}U^2\cos(2\omega t) + \frac{a_3}{4}U^3\cos(3\omega t) + \dots \end{aligned} \tag{A.20}$$

The second order harmonic distortion (HD$_2$) of this amplifier is defined as the strength of the second-order component relative to the first-order output signal component. The third-order harmonic distortion (HD3) is defined as the ratio between the third- and first-order components in the output signal (A.21):

$$\begin{aligned} HD_2 &= \frac{1}{2}\frac{a_2}{a_1}U \quad \text{(second-order harmonic distortion)} \\ HD_3 &= \frac{1}{4}\frac{a_3}{a_1}U^2 \quad \text{(third-order harmonic distortion)} \end{aligned} \tag{A.21}$$

As expected and confirmed by Formula (A.21), the level of distortion is also related to the amplitude U of the input signal. It follows that, independently of the amplifier configuration, distortion can always be suppressed by reducing the input signal level. On the other hand, the input signal is only one part of the story. The noise floor is also an important factor, since the total harmonic distortion plus noise (THD+N) is a limiting factor for the signal quality, especially for untuned wideband amplifiers where the noise floor is integrated over a wide frequency band.

SELF-MIXING COMPONENTS AND DC-OFFSET

Also interesting is the unexpected offset component in the output signal (underlined in A.21). Usually the term $a_2/2\ U^2$ can be neglected. For large input signals however, this term causes an unwanted DC-offset at the output of the amplifier.

For example, in an frequency conversion stage where a large LO-signal is applied to the input of the mixer, a serious DC-offset may appear at the output of the mixer. In a direct conversion receiver, where the RF-signal is converted immediately to a lower frequency band around DC, this results in the irreversible loss of near-DC signal information, because it is very difficult to separate the unwanted offset signal and the signal-of-interest in a nearby frequency band.

The frequency independent time-domain model of the nonlinear amplifier in Equation (A.19) can be embedded in a negative feedback loop, as is illustrated by Figure A.6. As a result, the nonlinear characteristics of the active element $a(x)$ in the forward path of the loop are suppressed in the overall transfer characteristic. The closed-loop representation of this system can again be reduced to a polynomial expression of the form (A.22):

$$y(v_{in}) = b_0 + b_1 v_{in} + b_2 v_{in}^2 + b_3 v_{in}^3 + \dots, \tag{A.22}$$

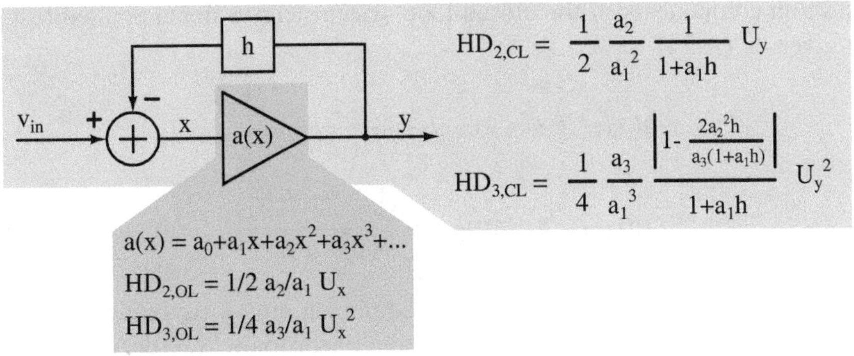

Figure A.6. Distortion in a basic feedback amplifier with a frequency independent nonlinear active element in the forward path. Note that the level of distortion suppression of the feedback system is (in first order approximation) inverse proportional to the feedback factor h.

where the coefficients $b_0 \dots b_3$ can be approximated by expressions (A.23):

$$b_0 = \frac{a_0}{1+a_1h} \quad \text{(A.23a)}$$

$$b_1 = \frac{a_1}{1+a_1h} \quad \text{(A.23b)}$$

$$b_2 = \frac{a_2}{(1+a_1h)^3} \quad \text{(A.23c)}$$

$$b_3 = \frac{a_3(1+a_1h) - 2a_2^2h}{(1+a_1h)^5} \quad \text{(A.23d)}$$

From the DC-coefficient b_0 of Equation (A.23a), it is easy to understand that the offset voltage of the active element is suppressed by the first order loop gain a_1h. It is also well-understood that the first-order gain b_1 of the closed-loop system (A.23b) reduces to the expression $1/h$ if a sufficiently large excess gain (a_1h) is available in the loop. On the other hand, it is a little bit less evident to grasp why the second- and third-order coefficients of the closed-loop polynomial decline with the third and the fourth power of a_1h, respectively. In order to simplify matters, the problem is split in two stages.

As a start, by using the definitions of the second- and third-order harmonics introduced in Equation (A.21), it is fairly straightforward to find out that the distortion components of the closed-loop system with a signal applied to v_{in} are given by (A.24):

$$\begin{aligned} \mathrm{HD}_{2,\mathrm{CL}} &\approx \mathrm{HD}_{2,\mathrm{OL}} \cdot \frac{1}{(1+a_1h)^2} \\ \mathrm{HD}_{3,\mathrm{CL}} &\approx \mathrm{HD}_{3,\mathrm{OL}} \cdot \frac{\left|1-\frac{2a_2^2h}{a_3(1+a_1h)}\right|}{(1+a_1h)^3}, \end{aligned} \tag{A.24}$$

from which it is clear that distortion component $\mathrm{HD}_{2,\mathrm{CL}}$ is reduced by the square of the loop gain, while the third order distortion $\mathrm{HD}_{3,\mathrm{CL}}$ decreases with the third-order of the loop gain. These observations can be easily verified by the examination of Figure A.6. Suppose both the input signal level v_{in} and the linear gain ($b_1 \approx 1/h$) of the closed-loop system remain constant. As a consequence, also the amplitude at node y remains fixed. When the gain a_1 of the active element is increased, it follows that the signal amplitude at input node x of the active element must change inverse proportional to a_1. Now, keeping the connection between forward gain a_1 and the input signal amplitude at node x in mind, and from the original definition of the open-loop distortion (A.21), an intuitive validation for the correctness of closed-loop Equation (A.24) is obtained.

For a truly fair comparison between the harmonic distortion of the open-loop and closed-loop amplifier, the expressions of (A.24) should referred to the output amplitude U_{o} of the system (A.25):

$$\begin{aligned} \mathrm{HD}_{2,\mathrm{CL}} &\approx \frac{1}{2}\frac{a_2}{a_1^2}\,\frac{1}{1+a_1h}\,U_{\mathrm{o}} \\ \mathrm{HD}_{3,\mathrm{CL}} &\approx \frac{1}{4}\frac{a_3}{a_1^3}\,\frac{\left|1-\frac{2a_2^2h}{a_3(1+a_1h)}\right|}{1+a_1h}\,U_{\mathrm{o}}^2 \end{aligned} \tag{A.25}$$

Remark that if the feedback factor h is reduced to 0, the expressions for the open-loop system are obtained. A particularly interesting result here is that for a certain amplifier (with all a_i fixed), the gain of the closed-loop system, controlled by h, can be linearly exchanged with second-order distortion suppression performance. For the $\mathrm{HD}_{3,\mathrm{CL}}$ third-order component, the net result is even better than linear due to the reduction in the numerator in Formula (A.25).

For small values of U_{o}, it might be tempting to conclude that third-order distortion in the output of the amplifier is only a negligible side effect compared to the larger second-order distortion component. The opposite turns out to be true if the amplifier is being used in the analog front-end of a heterodyne radio receiver. For example, suppose a strong 475 MHz blocker signal is present at the receiving antenna of a 950 MHz GSM front-end. Second-order nonlinearities in the input stage of the receiver will generate undesired in-band spurious frequencies. Fortunately, because these blocker signals are far beyond the passband of the receiver, they can be effectively suppressed by the band-select filter. For this purpose, the RF input stage of a GSM receiver is commonly equipped with a passive 35 MHz-wide SAW-filter, which can provide over 40 dB out-of-band attenuation at large offsets from the passband frequency.

It turns out that third-order nonlinearities are a far more severe issue due to their intermodulation beat products between in-band blocker signals. Suppose two large interferers at closely separated frequencies ω_{b1} and ω_{b2} are present in the input signal. Third-order intermodulation ($\text{IM}_3 = 3\text{HD}_3$) generates unexpected distortion components at $2\omega_{b1} - \omega_{b2}$ and $2\omega_{b2} - \omega_{b1}$, which are located closely to the frequency of the blocker signals [San99]. Unfortunately, those extra intermodulation components can corrupt a more distant and weaker signal that could be of interest to the user. Since all signals are within the passband of the band-selection filter, an irreversible loss of the weak signal may arise from third-order harmonic distortion in the analog receive chain.

Harmonic distortion specifications of a GSM receiver

As a practical example, consider the GSM 05.05 radio specification. In this standard, it is defined which levels of blocking signals a handheld GSM device must be able to withstand without dropping a call. The in-band blocking signal level that the standard imposes for a mobile station (MS) defines the linearity requirements of the front-end.

At the same moment when two interferers with a power level of −49 dBm are present at the antenna terminal of the receiver, the sensitivity level of the receiver must be equal or better than −99 dBm. Since the baseband processor of a GSM requires a minimal carrier/noise-ratio of 8 dB for a successful detection of the GMSK signal, it is straightforward to find that the maximum tolerable intermodulation component is 8 dB below the sensitivity level, which is −107 dBm. The third-order intermodulation requirement of the front-end is defined by the ratio between the power of the beat product and the power of the interferers (A.25):

$$\begin{aligned} \text{IM}_3\ [\text{dB}] &= \text{beat power [dBV]} - \text{blocker level [dBV]} \\ &= \text{beat power [dBW]} - \text{blocker level [dBW]} \\ &= -107\text{dBm} + 49\text{dBm} = -58\text{dB} \end{aligned} \tag{A.26}$$

The linearity of the analog front-end of a receiver is commonly characterized by its third-order intermodulation intercept point (IP_3), being the interferer power level at which the distortion power becomes equal to the value of the signal-of-interest in the spectral density plot of the amplification chain of the receiver. If the interferer power is referred to the input of the system, this power level is called the input-referred third-order intermodulation intercept point (IIP3). It is called the output-referred IP3 if the interferer power is referred to the output node of the signal chain. The minimal value of the corresponding intermodulation intercept point IIP3 of a mobile GSM receiver (MS) can now be obtained by extrapolating the blocker level until IM3 becomes equal to 0 dB, which is exactly the point where the third-order distortion beat power at the output of the amplifier becomes equal to the power of the signal-of-interest (A.26):

$$\begin{aligned} \text{IIP}_3\,[\text{dBW}] &= \text{blocker level [dBW]} - \frac{1}{2}\,\text{IM}_3\text{@ blocker level [dB]} \\ &= -49\text{dBm} + \frac{1}{2}\,58\text{dB} = -20\text{dBm} \end{aligned} \quad (A.27)$$

Distortion in a single-stage MOS amplifier

To illustrate the usefulness of the formulas that were introduced in the previous section, the distortion components of a single MOS transistor amplifier with resistive source degeneration are calculated. The quadratic transistor model of (A.3) is used, but the calculations can easily be expanded to any other transistor model. It will also turn out that degeneration causes third-order harmonic distortion components, although only second-order coefficients are taken into account in this simplified transistor model. The transistor schematic with resistive degeneration in the feedback path and the corresponding small-signal model are represented by Figure A.7.

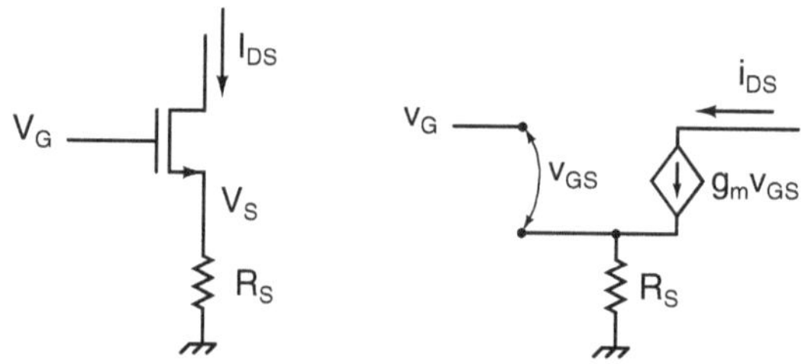

Figure A.7. A MOS transistor amplification stage with resistive degeneration in the feedback path. The voltage drop over the resistor reduced the overdrive voltage V_{GST} over the transistor. This reduces the overall gain of the circuit, but will also suppress second-order distortion components.

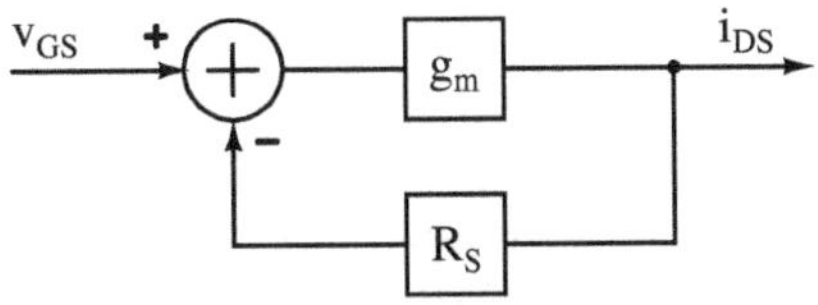

Figure A.8. Simplified equivalent block diagram of a MOS transistor with resistive degeneration. Note that parasitic effects and the output impedance of the transistor are not taken into account.

The feedback part in this topology is taken into account by the series resistor R_S which is placed in the output current path at the source of the transistor. The output current I_{DS} causes a current dependent voltage drop over R_S. The negative feedback to the input of this system is thus realized by a decrease of the effective gate-source overdrive voltage V_{GST} that is applied to the transistor. The resulting closed-loop transconductance gain G_m of this circuit is given by (A.28):

$$G_m = \frac{g_m}{1 + g_m R_S} \tag{A.28}$$

The circuit of Figure A.7 can also be represented by using the familiar feedback block diagram of Figure A.8. Of course, in order to study the distortion characteristics of the degenerated transistor stage, a more realistic representation of the transconductance characteristics of the active element is required. For this reason, the small-signal gain g_m in the forward path is replaced by a more accurate and large-signal model of the voltage-to-current gain of a MOS transistor. Starting from the quadratic MOS transistor model of Equation (A.3), the input signal of the transistor is split in a DC-component (capital letters) and it's superimposed smaller AC-signal (lower case) (A.29):

$$\begin{aligned}
I_{\mathrm{DS}} + i_{\mathrm{DS}} &= K'\frac{W}{L}\left(V_{\mathrm{GST}} + v_{\mathrm{G}}\right)^2 \\
&= K'\frac{W}{L}\left(V_{\mathrm{GST}}^2 + 2V_{\mathrm{GST}}\, v_{\mathrm{G}} + v_{\mathrm{G}}^2\right) \\
i_{\mathrm{DS}} &= K'\frac{W}{L}\left(2V_{\mathrm{GST}} v_{\mathrm{G}} + v_{\mathrm{G}}^2\right) \\
i_{\mathrm{DS}} &= \underbrace{g_m}_{a_1}\, v_{\mathrm{G}} + \underbrace{\frac{g_m}{2V_{\mathrm{GST}}}}_{a_2}\, v_{\mathrm{G}}^2
\end{aligned} \tag{A.29}$$

Expanding this equation results in a static part, representing the biasing point of the transistor and a linear part, representing the small-signal gain g_m. Also, a smaller second-order component is present. It is this factor that will contribute to both the second- and third-order distortion components when the transistor is embedded in the resistive degenerated feedback system. Based on Formula (A.29), one could conclude that an increase of the overdrive voltage V_{GST} is the correct way to tackle distortion. The reality, which is of course a little bit more complex, requires a more subtle shaded answer, depending on some boundary conditions. Based on the closed-loop distortion formula (A.24) and from the polynomial coefficients of the MOS transistor (A.29), the second-order distortion performance of the degenerated MOS stage can be approximated by (A.30):

$$\mathrm{HD}_{2,\mathrm{CL}} = \frac{1}{4} \underbrace{\frac{v_{\mathrm{G,peak}}}{V_{\mathrm{GS}} - V_T}}_{\text{relative swing}} \frac{1}{(1 + \underbrace{g_m R_S}_{\text{loop gain}})^2}, \tag{A.30}$$

from which it is clear that the distortion at the output of the amplifier is dependent on the relative voltage swing (w.r.t. the overdrive voltage) at the input and the loop gain $g_m R_S$. In a discrete amplifier setup, where fixed transistor dimensions (W/L) are a constraint, increasing the overdrive voltage V_{GST} or the series resistance R_S are appropriate choices to improve the linearity of the single transistor amplifier.

The calculations for the harmonic distortion can also be referred to the current amplitude i_{DS} at the output of the system. A combination of the MOS transistor parameters from (A.29), the output-referred second-order distortion formula in Equation (A.25) and some unpleasant calculations may lead the reader to the following expression for the output-referred second-order distortion (A.31):

$$\mathrm{HD}_{2,\mathrm{CL}} = \frac{1}{8} \underbrace{\frac{i_{\mathrm{DS,peak}}}{I_{\mathrm{DS}}}}_{\text{relative swing}} \frac{1}{1 + \underbrace{g_m R_S}_{\text{loop gain}}} \tag{A.31}$$

Then, keeping the closed-loop transconductance of the overall system in mind (Formula (A.28)), the output-referred distortion from above can also be reformulated as (A.32):

$$\mathrm{HD}_{2,\mathrm{CL}} = \frac{1}{8} \frac{i_{\mathrm{DS,PEAK}}}{I_{\mathrm{DS}}} \left(\frac{G_m}{g_m} \right) \tag{A.32}$$

The last factor in this expression indicates the ratio between the closed-loop transconductance G_m and the open-loop gain g_m that is available from the transistor. It is thus again confirmed by this observation that gain can be traded for linearity in a feedback system. Using the approximation of g_m introduced by

(A.3), the second-order distortion can also be expressed in terms of more practical design parameters such as the closed-loop transconductance G_m, the bias current I_{DS} and the overdrive voltage V_{GST} (A.33):

$$\mathrm{HD}_{2,\mathrm{CL}} = \frac{1}{8}\frac{i_{\mathrm{DS,peak}}}{I_{\mathrm{DS}}}\frac{(V_{\mathrm{GS}} - V_T)}{2\,I_{\mathrm{DS}}}\,G_m \tag{A.33}$$

The closed-loop transconductance G_m represents the useful gain that is eventually available from this system. Furthermore, the bias current I_{DS} will define the power consumption of the amplifier setup. Note that if the power consumption is a fixed constraint, the linearity can still be improved by *decreasing* the overdrive voltage, completely opposite to the intuitive solution suggested earlier. The reason for this behaviour is that a constant drain-source current leads to a quadratical increase of the dimensions of the transistor (W/L), thereby also increasing the small-signal gain g_m. A major pitfall of improving the linearity in this way is that the high-frequency performance of the system will collapse under the increased load of parasitic capacitances.

The simplified MOS transistor model used in these calculations does not contain any third-order coefficients. When the transistor is used in an open-loop topology, no third-order distortion components will thus appear at the output. However, from Equation (A.24) it is clear that third-order components appear in the degenerated amplifier setup after all. By following the same strategy as for the second-order distortion, and employing the polynomial coefficients from (A.29) in $\mathrm{HD}_{3,\mathrm{CL}}$, the following expressions are obtained for the third-order harmonic distortion (A.34):

$$\begin{aligned}\mathrm{HD}_{3,\mathrm{CL}} &= \frac{1}{32}\frac{v^2_{\mathrm{G,peak}}}{(V_{\mathrm{GS}} - V_T)^2}\frac{g_m R_S}{(1 + g_m R_S)^4} \\ &= \frac{1}{32}\left(\frac{i_{\mathrm{DS,peak}}}{I_{\mathrm{DS}}}\right)^2\left(\frac{G_m}{g_m}\right)^2 g_m R_S \end{aligned}\tag{A.34}$$

While these formulas might look very impressive or confusing for the reader, it is far more interesting to look back to the closed-loop second-order harmonic distortion formula of A.32. It seems that the third-order distortion of A.34 is related to $\mathrm{HD}_{2,\mathrm{CL}}$ in the following manner (A.35):

$$\begin{aligned}\mathrm{HD}_{3,\mathrm{CL}} &= 2\left[\frac{1}{8}\frac{i_{\mathrm{DS,peak}}}{I_{\mathrm{DS}}}\left(\frac{G_m}{g_m}\right)\right]^2 g_m R_S \\ &= 2\left(\mathrm{HD}_{2,\mathrm{CL}}\right)^2 g_m R_S \\ &= g_m\,\mathrm{HD}_{2,\mathrm{CL}}\;R_S\;2\mathrm{HD}_{2,\mathrm{CL}} \\ &= \underbrace{g_m\,\mathrm{HD}_{2,\mathrm{CL}}}_{(1)}\;\underbrace{R_S}_{(2)}\;\underbrace{\mathrm{IM}_{2,\mathrm{CL}}}_{(3)}\end{aligned}\tag{A.35}$$

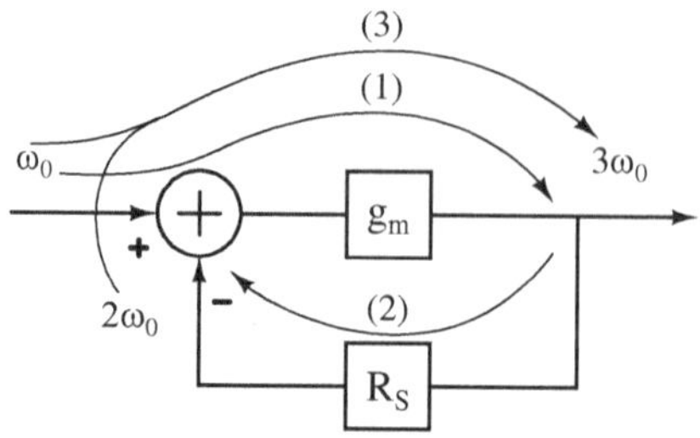

Figure A.9. The origins of third-order distortion in a MOS transistor amplifier with resistive degeneration. Second-order distortion components (1) appear at the output, reenters into the input of the system (2) and finally cause third-order distortion which is the result of the intermodulation mix product between the input signal and the second-order distortion components.

The latter method of formulating HD_3 provides a better insight in the distortion generating mechanism of the resistive degenerated amplifier. The amplitude of the second-order component in the output current i_{DS} is taken into account by the first factor (1) in Equation (A.35). This small-signal current is transformed to a voltage by R_S (2) which is subsequently fed back to the input of the system. When re-entering the loop, this second-order distortion component will be mixed with the fundamental input signal (single tone at ω_0). *Second-order* intermodulation distortion (IM_2), represented by factor (3), will finally generate frequency components at $2\omega_0 \pm \omega_0$. The sum of these components finally results in the third-order distortion. The whole process is once again illustrated in Figure A.9.

A.2 Frequency dependent distortion in feedback systems

The linearity calculations in the previous section have assumed that the active element in the forward path of the feedback loop has a frequency-independent characteristic. All real-world amplifier implementations however, exhibit one or more poles beyond which the gain is reduced. The loop-gain of the feedback system embedding such a frequency-limited amplifier is dependent on the operating frequency and as a result, also the distortion suppression capabilities will reduce at the higher end of the spectrum. Calculations of the distortion parameters of such a nonlinear closed-loop system are not straightforward. For linear time-invariant systems, the frequency dependent behaviour of the feedback system is easily described in the frequency domain. But for nonlinear systems, a complete description in terms of interdependent eigenvectors (i.e. a set of sinusoidal in- and outputs) is not possible. Rather than elaborating on a thorough mathematical description of the frequency dependent behaviour of nonlinear systems, a lot of insight in the internals of a feedback system can already be obtained by some well-considered reasoning.

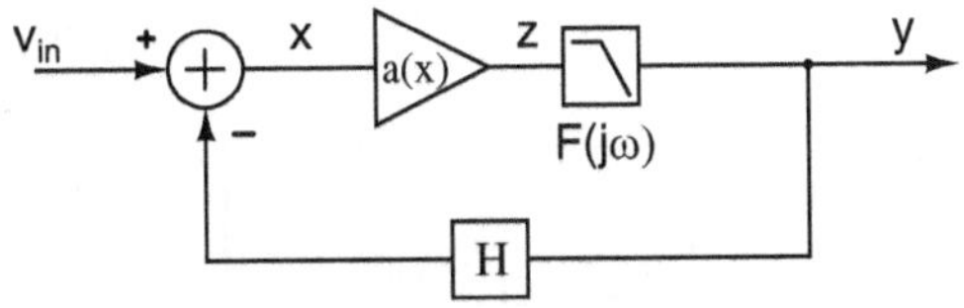

Figure A.10. Closed-loop system embedding a single-pole amplifier. Remark that the pole of the active element is located at its output. For other configurations, the frequency dependent distortion calculations should be adjusted accordingly.

To start with simplicity, suppose that the active element in the forward path can be split in a frequency dependent part $F(j\omega)$ and a nonlinear part $a(x)$ as is shown in Figure A.10. If only the first-order linear gain coefficient a_1 of the amplifier is considered, the problem is reduced to an LTI-system of which the transfer function can be easily determined. Apart from the closed-loop LTI distortionless transfer function from above, two extra transfer characteristics are important for the derivation of the frequency dependent distortion behaviour. The transfer characteristic from the input signal v_{in} to the signal at input node x of the active element is given by (A.36):

$$\text{TF}(j\omega)_{v_{in}\to x} = \frac{1}{1 + F(j\omega)a_1 H} \tag{A.36}$$

It follows that the amplitude at node x is reduced by the loop-gain of the system. A reduction of the signal swing at the input node of the active element has a large impact on the distortion produced by the amplifier. For example, consider the case of a single-pole amplifier with a pole at frequency ω_0. The transfer function from v_{in} to node x will show a zero at $\omega_z = \omega_{p1}$ and a pole at $\omega_p = \omega_{p1} a_1 H$ (A.37):

$$\begin{aligned} F(j\omega) &= \frac{1}{1 + j\omega/\omega_{p1}} \\ \text{TF}(j\omega)_{v_{in}\to x} &= \frac{1}{1 + a_1 H} \frac{1 + j\omega/\omega_{p1}}{1 + j\omega/[\omega_{p1}(1 + a_1 H)]} \end{aligned} \tag{A.37}$$

For a large loop gain and low frequencies, the linearity of the closed-loop system is thus better than that of the amplifier in an open-loop setup. Unfortunately, at the first zero in the open-loop transfer characteristic, the excess gain in the loop is gradually reduced, resulting in a degradation of the linearity performance of the closed-loop system. When the excess loop gain eventually drops to 0, it follows from Equation (A.36) that the amplitude at node x is equal to the input amplitude. At this point, the linearity will be no better than this of the open-loop amplifier. The entire process is visualized in Figure A.11, along

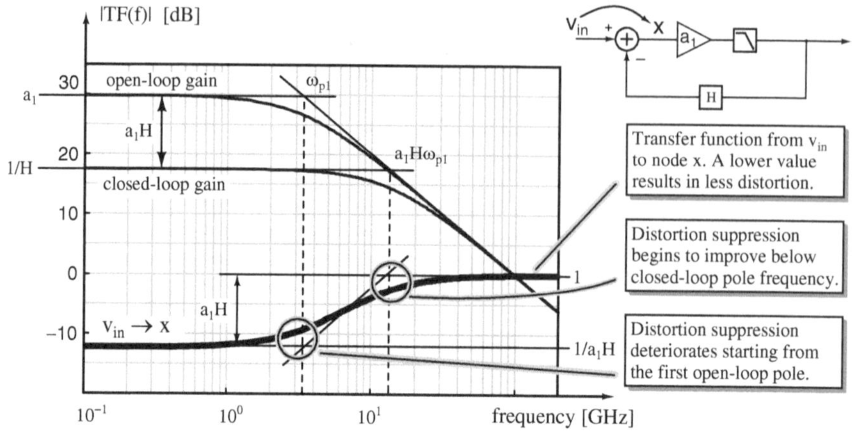

Figure A.11. Transfer function from v_{in} to intermediate node x for a single-pole amplifier. Distortion suppression is only available below the closed-loop pole of the system, from where the excess loop gain reduces the signal level at node x.

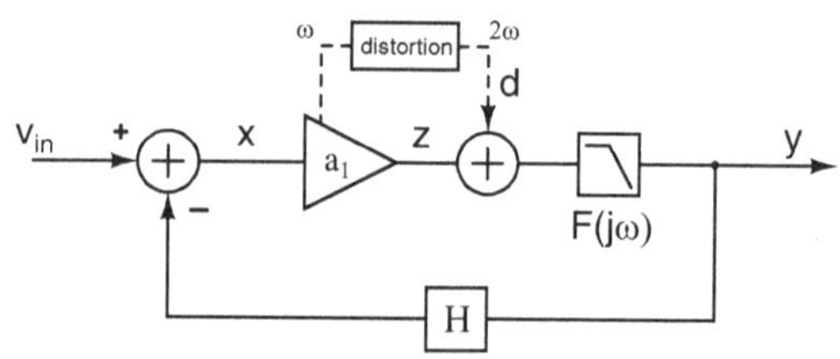

Figure A.12. The principle of distortion injection in a closed-loop system. From the input signal at node x, the distortion amplitude at node d can be calculated. The final distortion level at output node y is then found using the transfer function from d to y.

with the most important frequency marks. Also, this figure clearly illustrates once more that gain may be traded for linearity. But as soon as the available excess gain ($a_1 H$) drops due to the limited bandwidth of the active element, the linearity performance starts to degrade.

Starting from the transfer function from input terminal v_{in} to node x at the input of the active element, the amplitude of the fundamental component at the output of the amplifier (node z in Figure A.12) is found by a multiplication with the linear gain a_1 of the amplifier. In an open-loop setup, the magnitude of the harmonic components would then easily be determined using the $\text{HD}_{2,\text{OL}}$ and $\text{HD}_{3,\text{OL}}$ formulas of (A.21). In the closed-loop system however, distortion must be regarded as a brand new signal component that is injected in the system, as illustrated by Figure A.12. The second- or third order distortion signal components are subsequently transferred to the output of the closed-loop system, which is described by the transfer characteristic from node d to node y (A.38):

$$\mathrm{TF}(j\omega)_{d\to y} = \frac{F(j\omega)}{1 + F(j\omega)a_1 H} \tag{A.38}$$

Continuing the example of the single-pole amplifier, the transfer function of the injected distortion to the output shows a single pole at the cut-off frequency of the closed-loop system (A.39):

$$\begin{aligned} F(j\omega) &= \frac{1}{1 + j\omega/\omega_{p1}} \\ \mathrm{TF}(j\omega)_{z\to y} &= \frac{1}{1 + a_1 H}\,\frac{1}{1 + j\omega/[\omega_{p1}(1 + a_1 H)]} \end{aligned} \tag{A.39}$$

IMPORTANT REMARK

It is of vital importance to recognize that, starting from the point where distortion is injected at node z in the loop, the transfer function to the output should be evaluated at the particular frequency of that distortion component.

For example, if the fundamental frequency of the input signal is ω_{fund}, nonlinearities will produce a second-order harmonic spur at $2\omega_{\mathrm{fund}}$ which is injected at node d. From that moment on, all calculations must be performed at frequency $2\omega_{\mathrm{fund}}$. The frequency conversion makes the modelling of the system with a transfer function unfeasible, without falling back on more complicated algorithms such as the harmonic transfer matrices (HTM) method.

Second-order frequency dependent distortion

Putting all pieces together, the calculation of frequency dependent distortion in the closed-loop system is obtained in a two-step approach. In order not to overcomplicate matters, the calculation for HD_2 is performed first, then followed by the slightly more complex HD_3 derivation. Using Figure A.12, the first step is to determine the amplitude of the fundamental frequency at node z, which is the output of the active gain element. The amplitude is easily determined using the transfer function from the input to node x (A.36), multiplied by the first-order linear gain a_1 (A.40):

$$\mathrm{FUND}_z(j\omega) = v_{in} \cdot \frac{1}{1 + F(j\omega)a_1 H} \cdot a_1 \tag{A.40}$$

The second-order distortion characteristic (HD$_2$) of the active gain element when it is embedded in the forward path of a feedback loop is dependent on the signal level applied to its input (node x in Figure A.12). The transfer function from the input v_{in} to node x is a function of frequency and is given by (A.36). With the original open-loop distortion formula (A.21) in mind, the frequency dependent closed-loop harmonic distortion ratio at node z is given by (A.41):

$$\begin{aligned} \mathrm{HD}_{2,\,\mathrm{z}}(j\omega) &= \mathrm{HD}_{2,\,\mathrm{OL}} \cdot \mathrm{TF}(j\omega)_{v_{in}\rightarrow x} \\ &= \mathrm{HD}_{2,\,\mathrm{OL}} \cdot \frac{1}{1+F(j\omega)a_1 H} \end{aligned} \tag{A.41}$$

Finally, by combining the amplitude of the fundamental frequency of (A.40) and the closed-loop harmonic distortion formula of (A.41), the absolute amplitude of the second-order harmonic component at node z is obtained (A.42):

$$\begin{aligned} \mathrm{HARM}_{2,z}(j2\omega) &= \mathrm{FUND}_z(j\omega) \cdot \mathrm{HD}_{2,\,\mathrm{z}}(j\omega) \\ &= \mathrm{HD}_{2,\,\mathrm{OL}} \cdot \left[\frac{1}{1+F(j\omega)a_1 H}\right]^2 \cdot a_1 \cdot v_{in} \end{aligned} \tag{A.42}$$

As explained before in Figure A.12, the above second-order harmonic signal component is reinjected in the feedback loop at node d, before it finally appears at the output node of the closed-loop system. The transfer function from d to output y is described by (A.38). It must be emphasized that from this point on, all transfer expressions must be evaluated at $2\omega_{\mathrm{fund}}$, which is the frequency of the second-order harmonic (A.43):

$$\begin{aligned} \mathrm{HARM}_{2,y}(j2\omega) &= \mathrm{HARM}_{2,z}(j2\omega) \cdot \mathrm{TF}(j2\omega)_{d\rightarrow y} \qquad \text{(A.43)} \\ &= \mathrm{HD}_{2,\,\mathrm{OL}} \cdot \left[\frac{1}{1+F(j\omega)a_1 H}\right]^2 \cdot \frac{F(j2\omega)}{1+F(j2\omega)a_1 H} \cdot a_1 \cdot v_{in} \end{aligned}$$

In order to obtain a closed form expression for the closed-loop frequency dependent harmonic distortion, the fundamental frequency component at the output of the loop must be determined (A.44):

$$\mathrm{FUND}_y(j\omega) = \frac{F(j\omega)a_1}{1+F(j\omega)a_1 H} \cdot v_{in} \tag{A.44}$$

Ultimately, the closed-loop distortion expression is now easily found by the ratio of the second-order harmonic $\mathrm{HARM}_{2,y}(j2\omega)$ to the fundamental component $\mathrm{FUND}_y(j\omega)$ (A.45):

CLOSED-LOOP SECOND-ORDER FREQUENCY DEPENDENT DISTORTION

$$\mathrm{HD}_{2,\,\mathrm{CL}}(j\omega) = \frac{\mathrm{HARM}_{2,y}(j2\omega)}{\mathrm{FUND}_y(j\omega)} \tag{A.45}$$

$$= \mathrm{HD}_{2,\,\mathrm{OL}} \cdot \underbrace{\frac{1}{1+F(j\omega)a_1H}}_{(1)\ \mathrm{TF}(j\omega)_{v_{in}\to x}} \cdot \underbrace{\frac{F(j2\omega)}{1+F(j2\omega)a_1H}}_{(2)\ \mathrm{TF}(j2\omega)_{d\to y}} \cdot \underbrace{\frac{1}{F(j\omega)}}_{(3)\ \frac{\mathrm{FUND}_z(j\omega)}{\mathrm{FUND}_y(j\omega)}}$$

Note that when the $F(j\omega)$ filter characteristic is chosen equal to unity, exactly the same result as for the frequency-independent system in (A.24) is obtained. Furthermore, three separate factors play an important role in the distortion characteristic of (A.45). The first factor (A.45-1) is the transfer characteristic from the input signal v_{in} to the input node x of the amplifier. At low frequencies, below the first pole of $F(j\omega)$, it follows that the signal swing at node x is actively suppressed by the loop gain of the system. Secondly (A.45-2), the closed loop system under consideration provides an additional level of distortion suppression: harmonic components that still emerge at the output of the amplifier are additionally suppressed by the transfer function from node d to node y before they appear at the output of the system. Finally, it is important to recognize that also the fundamental frequency component plays a role in the harmonic distortion expression. The last factor (A.45-3), takes into account that the fundamental frequency component is affected by the $F(j\omega)$ filter characteristic.

$$F(j\omega) = \frac{1}{1+j\omega/\omega_{p1}} \tag{A.46}$$

$$\mathrm{HD}_{2,\,\mathrm{CL}}(j\omega) = \frac{\mathrm{HD}_{2,\,\mathrm{OL}}}{(1+a_1H)^2} \cdot \frac{1+j\omega/\omega_{p1}}{1+j\omega/\left[\omega_{p1}(1+a_1H)\right]} \cdot \frac{1+j\omega/\omega_{p1}}{1+j\omega/\left[\omega_{p1}\frac{(1+a_1H)}{2}\right]}$$

When the role of the general filter characteristic $F(j\omega)$ is filled in by an elementary single-pole filter (A.46), the second-order distortion characteristic shows a double zero at the filter's cut-off frequency ω_{p1}. As a consequence, the linearity performance of the closed-loop system starts to *degrade* at a rate of 20 dB/dec for frequencies beyond this pole. The $\mathrm{HD}_{2,\,\mathrm{CL}}$ characteristic also exhibits two non-coinciding poles. One pole is located at half the cut-off frequency of the closed loop system ($\frac{1}{2}a_1H\omega_{p1}$) and is due to the fact that second-order harmonic components have twice the fundamental frequency. As a result, the corner frequency of the transfer function from node d to the output

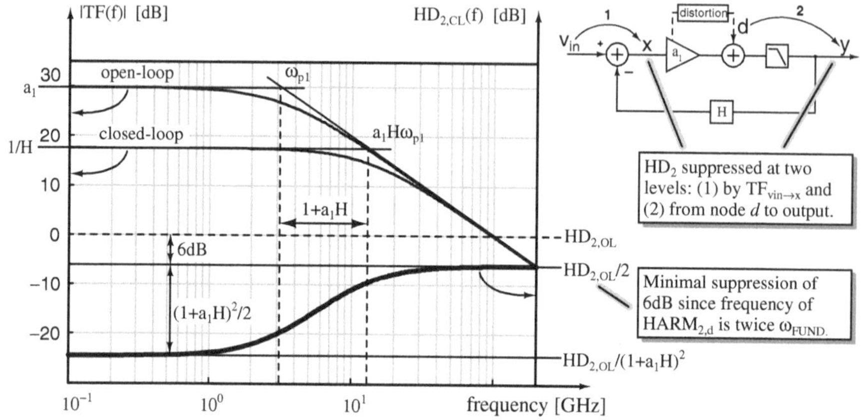

Figure A.13. Second-order distortion suppression in a feedback system. For frequencies below the first pole of the active element (ω_{p1}), distortion is suppressed by the square of the excess gain. Beyond this frequency however, the $\mathrm{HD}_{2,\mathrm{CL}}$ suppression degrades at a rate of 20 dB/dec.

is reached somewhat earlier in case of the second harmonic. Likewise, the second pole ($a_1 H\omega_{p1}$) is caused by the pole in the transfer function from the input to node x, which must be evaluated at the frequency of the fundamental component.

Figure A.13 summarizes all the conclusions. In the single-pole feedback loop of Figure A.10, the open-loop distortion performance $\mathrm{HD}_{2,\,\mathrm{OL}}$ of the amplifier is improved by a factor dependent on the loop gain. Starting from the first pole of the amplifier however, the excess loop gain starts to reduce as a result of which the distortion performance is also decreased at a rate of approximately 20 dB/dec. Finally, when the first pole of the closed-loop system is reached, the contribution of the feedback loop to the improvement of the linearity characteristic of the active gain element in the forward path becomes negligible.

Third-order frequency dependent distortion

Third-order distortion components in a closed-loop system mainly originate from only two sources. The first source, described in this section, is brought into existence by the third-order coefficient a_3 in the power series approximation of the amplifier. Using the same approach as was elaborately unfold in the previous section, the closed-loop version of the third-order distortion can – after some calculation which is omitted here – be expressed as follows (A.47):

CLOSED-LOOP THIRD-ORDER FREQUENCY DEPENDENT DISTORTION PART 1/2

$$\mathrm{HD}_{3,\,\mathrm{CL}}(j\omega) = \frac{\mathrm{HARM}_{3,y}(j3\omega)}{\mathrm{FUND}_y(j\omega)} \tag{A.47}$$

$$= \mathrm{HD}_{3,\,\mathrm{OL}} \cdot \underbrace{\left[\frac{1}{1+F(j\omega)a_1H}\right]^2}_{(1)\ \mathrm{TF}(j\omega)^2_{v_{in}\to x}} \cdot \underbrace{\frac{F(j3\omega)}{1+F(j3\omega)a_1H}}_{(2)\ \mathrm{TF}(j3\omega)_{d\to y}} \cdot \underbrace{\frac{1}{F(j\omega)}}_{(3)\ \frac{\mathrm{FUND}_Z(j\omega)}{\mathrm{FUND}_Y(j\omega)}}$$

This time, the only noticeable difference with the HD_2 calculations is the quadratic dependency of HD_3 on the input signal level. This is taken into account by the square of the transfer function from the input of the system v_{in} to the input node x of the gain stage. Intuitively, one can predict that the extra factor of distortion suppression will only be of any significant importance in the lower frequency region of the transfer function. Starting from the first pole ω_{p1} in the open-loop system, third-order harmonic suppression will degrade much faster than it is the case for HD_2. This effect can easily be verified if $F(j\omega)$ in the generic formula of (A.47) is replaced by a first-order filter. Three concurrent zeroes appear at the 3 dB cut-off frequency ω_{p1} of the gain stage in the forward path (A.48):

$$F(j\omega) = \frac{1}{1+j\omega/\omega_{p1}} \tag{A.48}$$

$$\mathrm{HD}_{3,\mathrm{CL}}(j\omega) = \frac{\mathrm{HD}_{3,\mathrm{OL}}}{(1+a_1H)^3} \cdot \left[\frac{1+j\omega/\omega_{p1}}{1+j\omega/\left[\omega_{p1}(1+a_1H)\right]}\right]^2 \cdot \frac{1+j\omega/\omega_{p1}}{1+j\omega/\left[\omega_{p1}\frac{(1+a_1H)}{3}\right]}$$

The attentive reader may have noticed that there is a difference between frequency dependent harmonic distortion and frequency dependent intermodulation. Depending on the particular harmonic component under examination, Formula (A.47) should be adjusted to the correct frequency.

For example, in case of third-order distortion, the harmonic frequency component is located at three times the fundamental frequency. In Formula (A.47), this is already taken into account by the transfer function from node d to output node y, which is evaluated at three times the fundamental frequency. This makes that the expression for HD_3 is not entirely correct if employed to calculate third-order intermodulation distortion (IM_3). Suppose the input signal contains frequency components at ω_1 and at ω_2. Third-order intermodulation will generate spurious frequencies at both $2\omega_1 - \omega_2$ and $2\omega_2 - \omega_1$. If the two fundamental frequencies are close to each other, it follows that the spurs will also be located close to the fundamental frequency. As a result, the transfer function from node d to the output y of the system should thus be evaluated at the fundamental frequency. The expression for frequency dependent third-order intermodulation distortion is now given by (A.49):

THIRD-ORDER FREQUENCY DEPENDENT INTERMODULATION DISTORTION

$$\mathrm{IM}_{3,\,\mathrm{CL}}(j\omega) = 3 \cdot \mathrm{HD}_{3,\,\mathrm{OL}} \cdot \left[\frac{1}{1 + F(j\omega)a_1 H}\right]^3 \tag{A.49}$$

For the lower end of the frequency band, the approximation $\mathrm{IM}_3 = 3 \cdot \mathrm{HD}_3$ still holds. For higher frequencies, the HD_3 expression (A.47) exhibits a pole located at one third of the closed-loop pole. In case of intermodulation, this pole must be relocated to the cut-off frequency of the closed-loop system. For the same amplifier, the intermodulation performance will be 19 dB worse compared to the HD_3 characteristic at the higher end of the spectrum. In a practical application, if the bandwidth of the waveform that is applied to the input is kept below the cut-off frequency of the closed-loop system, the error on the approximation $\mathrm{IM}_3 = 3{\cdot}\mathrm{HD}_3$ is much less pronounced. Summarizing plot (Figure A.14) shows that this error can indeed be safely ignored. The reader should be aware

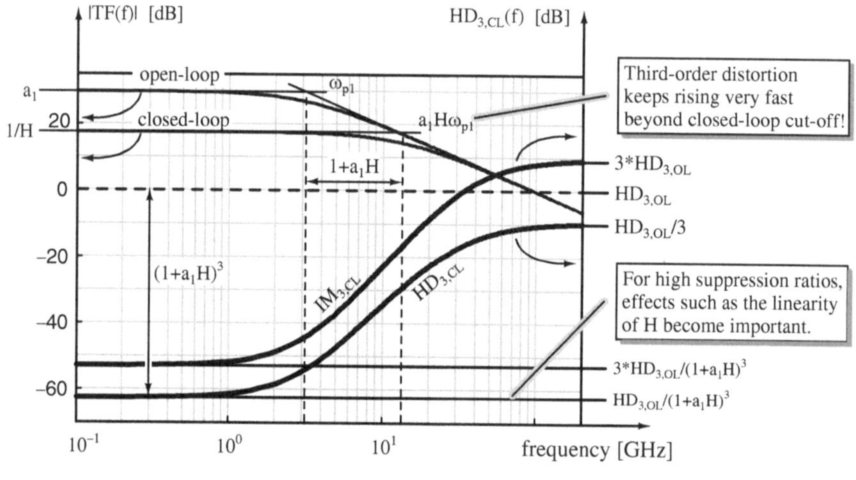

Figure A.14. Frequency dependent third-order harmonic distortion suppression in a feedback system. Note that only the effects originating from the third-order coefficient a_3 of the active element in the forward path were taken into account by this figure.

that only third-order effects originating from the third-order coefficient a_3 of the amplifier were taken into account during the creation of this figure.

Feedback-induced third-order distortion

In the previous section, third-order distortion originating from the third-order coefficient in the polynomial approximation of the nonlinear gain stage has been described. In Section A.1, it was illustrated that third-order harmonics can also emerge in a closed-loop system embedding an amplifier which only exhibits a second-order nonlinearity. This is because second-order harmonics are not too embarrassed to take a second turn in the loop: the harmonics are fed back to the input of the system and summed to the fundamental at the feedback point. During the second pass through the amplifier, *second-order* intermodulation products will cause spurs at three times the fundamental frequency. This process is illustrated in Figure A.15, from which the closed-loop $\text{HD}_{3,\,\text{CL}}(j\omega)$ expression can be found in only a few basic steps.

First, the amplitude of the second-order harmonic ($\text{HARM}_{2,d}$) at node d has to be determined. The exact derivation has already been shown in Section A.2, but the resulting equation is repeated below for the ease of our esteemed readers (A.50):

$$\text{HARM}_{2,d}(j2\omega) = \text{HD}_{2,\,\text{OL}} \cdot \left[\frac{1}{1 + F(j\omega)a_1 H}\right]^2 \cdot a_1\, v_{in} \tag{A.50}$$

The second-order harmonic component is reinjected at node d, takes a second turn in the loop, and eventually turns up again at the input x of the nonlinear gain stage. The amplitude of the second harmonic can be determined by using the transfer function from node d to the input x of the amplifier (A.51):

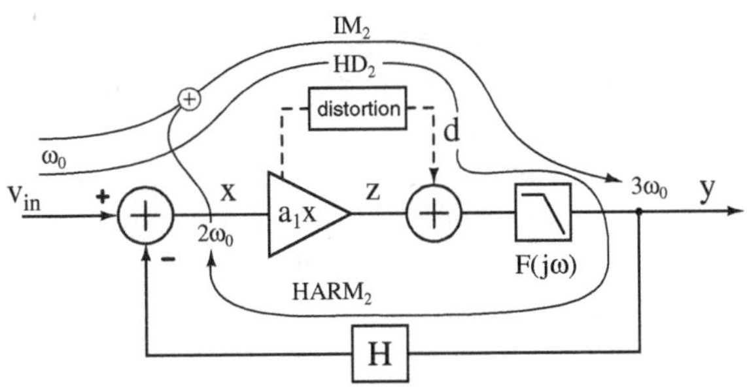

Figure A.15. Third-order harmonic distortion as a by-product of second-order intermodulation products between the input and second-order harmonics in the feedback path of a closed-loop system.

$$\begin{aligned}\mathrm{HARM}_{2,x}(j2\omega) &= \mathrm{HARM}_{2,d}(j2\omega)\cdot \mathrm{TF}(j2\omega)_{d\to x}\\ &= \mathrm{HARM}_{2,d}(j2\omega)\cdot \frac{F(j2\omega)H}{1+F(j2\omega)a_1 H}\end{aligned} \tag{A.51}$$

At this moment, two signals will be present at the input of the nonlinear amplification stage: the original fundamental component and the homesick second-order harmonic delivered by the feedback path. In a typical application, the amplitude of the fundamental component at node x is much larger than this of the second-order harmonic. As a consequence, the closed-loop intermodulation distortion characteristic $\mathrm{IM}_{2,z}$ is solely determined by the amplitude of the smaller second-order component.

At first glance, this statement may seem somewhat confusing, but can be easily clarified as follows. Consider the polynomial approximation of the amplifier where only the linear gain a_1 and the second-order coefficient a_2 are taken into account. The fundamental waveform with amplitude u_{fund} and the second-order harmonic with amplitude u_{harm2} are applied to the input of this amplifier (A.52):

$$\begin{aligned}a(x) &= a_1 x + a_2 x^2\\ &\qquad \text{with } x = u_{\mathrm{fund,x}}\cdot\cos(\omega t) + u_{\mathrm{harm2,x}}\cdot\cos(2\omega t)\\ a(x) &= \frac{a_2}{2}\left[u_{\mathrm{fund,x}}^2 + u_{\mathrm{harm2,x}}^2\right]\\ &\quad +\cos(\omega t)\cdot\left[a_1\, u_{\mathrm{fund,x}} + a_2\, u_{\mathrm{fund,x}}\, u_{\mathrm{harm2,x}}\right]\\ &\quad +\cos(2\omega t)\cdot\left[a_1\, u_{\mathrm{harm2,x}} + \frac{a_2}{2}\, u_{\mathrm{fund,x}}^2\right]\\ &\quad +\cos(3\omega t)\cdot\left[a_2\, u_{\mathrm{fund,x}}\, u_{\mathrm{harm2,x}}\right] + \cdots\end{aligned} \tag{A.52}$$

The output signal of the amplifier contains signals at various frequencies. However, only two of them are of importance for the calculation of the second-order intermodulation distortion at node z ($\mathrm{IM}_{2,z}$): the amplitude of the largest component at the fundamental frequency ω and the amplitude of the *second-order* intermodulation product at frequency $\omega_{\mathrm{fund}}+\omega_{\mathrm{harm2}} = 3\omega_{\mathrm{fund}}$. The ratio of the intermodulation product to the amplitude at the fundamental frequency provides the intermodulation distortion characteristic at node z of the amplifier. Remark that $\mathrm{IM}_{2,z}$ only depends on the amplitude (u_{harm2}) of the second-order distortion component (A.53):

$$\begin{aligned}\mathrm{IM}_{2,z} &= \frac{a_2\, u_{\mathrm{fund,x}}\, u_{\mathrm{harm2,x}}}{a_1\, u_{\mathrm{fund,x}} + a_2\, u_{\mathrm{fund,x}}\, u_{\mathrm{harm2,x}}}\\ &\approx \frac{a_2}{a_1}\cdot u_{\mathrm{harm2,x}}\end{aligned}$$

$$\begin{aligned} \mathrm{IM}_{2,\mathrm{z}}(j2\omega) &\approx \frac{a_2}{a_1} \cdot \mathrm{HARM}_{2,\mathrm{x}}(j2\omega) \\ &\approx \mathrm{IM}_{2,\mathrm{OL}} \cdot \frac{\mathrm{HARM}_{2,\mathrm{x}}(j2\omega)}{v_{in}} \end{aligned} \tag{A.53}$$

With the knowledge of the second-order intermodulation distortion ratio at node d, the effective amplitude of the intermodulation product at this node can be calculated. Subsequently, the value of the intermodulation product at the output of the system is found by using the transfer function from node d to the output y. Remember that the latter transfer characteristic should be evaluated at $j3\omega$, which is the beat frequency resulting from the second-order intermodulation product between the fundamental signal frequency and the second-order distortion component which is aimlessly wandering around the loop (A.54):

$$\begin{aligned} \mathrm{HARM}_{3,\mathrm{d}}(j3\omega) &= \mathrm{IM}_{2,\mathrm{z}}(j2\omega) \cdot \mathrm{FUND}_{\mathrm{z}}(j\omega) \\ \mathrm{HARM}_{3,\mathrm{y}}(j3\omega) &= \mathrm{IM}_{2,\mathrm{z}}(j2\omega) \cdot \mathrm{FUND}_{\mathrm{z}}(j\omega) \cdot \mathrm{TF}(j3\omega)_{d\rightarrow y} \\ &= \mathrm{IM}_{2,\mathrm{z}}(j2\omega) \cdot \frac{a_1\, v_{in}}{1+F(j\omega)a_1 H} \cdot \frac{F(j3\omega)}{1+F(j3\omega)a_1 H} \end{aligned} \tag{A.54}$$

The frequency dependent third-order distortion characteristic in (A.55) is defined as the ratio of the third-order harmonic component ($\mathrm{HARM}_{3,\mathrm{y}}$) to the amplitude of the fundamental component ($\mathrm{FUND}_{\mathrm{y}}$).

$$\begin{aligned} \mathrm{HD}_{3,\mathrm{CL}}(j\omega) &= \frac{\mathrm{HARM}_{3,\mathrm{y}}(j3\omega)}{\mathrm{FUND}_{\mathrm{y}}(j\omega)} \\ &= \mathrm{HARM}_{3,\mathrm{y}}(j3\omega) \cdot \frac{1+F(j\omega)a_1 H}{F(j\omega)\, a_1 v_{in}} \end{aligned} \tag{A.55}$$

Finally, the flattened expression for $\mathrm{HD}_{3,\mathrm{CL}}(j\omega)$ is deduced by combining Equations (A.50)–(A.55). The reader should keep in mind that the expression given below denotes the third-order distortion characteristic of a closed-loop system which includes an amplifier with only *second-order* nonlinearities. After some simplifications formula (A.56) is obtained:

CLOSED-LOOP THIRD-ORDER FREQUENCY DEPENDENT DISTORTION
PART 2/2

$$\begin{aligned} \mathrm{HD}_{3,\mathrm{CL}}(j\omega) &= \frac{\mathrm{HARM}_{3,\mathrm{y}}(j3\omega)}{\mathrm{FUND}_{\mathrm{y}}(j\omega)} \\ &= \frac{\mathrm{HD}_{2,\mathrm{OL}} \cdot \mathrm{IM}_{2,\mathrm{OL}} \cdot a_1}{[1+F(j\omega)a_1 H]^2} \cdot \underbrace{\frac{HF(j2\omega)}{1+F(j2\omega)a_1 H}}_{\mathrm{TF}(j2\omega)_{d\rightarrow x}} \cdot \underbrace{\frac{F(j3\omega)}{1+F(j3\omega)a_1 H}}_{\mathrm{TF}(j3\omega)_{d\rightarrow y}} \cdot \underbrace{\frac{1}{F(j\omega)}}_{\frac{\mathrm{FUND_z}(j\omega)}{\mathrm{FUND_y}(j\omega)}} \end{aligned} \tag{A.56}$$

The most noticeable difference with the $\mathrm{HD}_{3,\mathrm{CL}}$ distortion formula based on the a_3-coefficient in (A.47) is the appearance of an additional frequency dependent factor: $\mathrm{TF}(j2\omega)_{d\rightarrow x}$. Suppose $F(j\omega)$ is once again represented by a single-pole low-pass filter with cut-off frequency ω_{p1}, as shown in (A.57):

$$F(j\omega) = \frac{1}{1 + j\omega/\omega_{p1}} \tag{A.57}$$

$$\mathrm{HD}_{3,\mathrm{CL}}(j\omega) = \frac{2\mathrm{HD}_{2,\mathrm{OL}}^2 \cdot a_1}{(1+a_1H)^4} \cdot \frac{(1+j\omega/\omega_{p1})^3}{\left(1+\frac{j\omega}{\omega_{p1}(1+a_1H)}\right)^2} \cdot \underbrace{\frac{H}{1+\frac{j2\omega}{\omega_{p1}(1+a_1H)}}}_{\mathrm{TF}(j2\omega)_{d\rightarrow x}} \cdot \frac{1}{1+\frac{j3\omega}{\omega_{p1}(1+a_1H)}}$$

The extra transfer function will introduce a new pole located at $\omega_{p1}a_1H/2$. Intuitive reasoning on Figure A.15 learns that for increasing frequencies, the second-order harmonics is indeed increasingly suppressed by the filter, before being fed back to the input of the system. The net result is that the system contains three zeroes at frequency ω_{p1}, against four poles around the closed-loop cut-off frequency. As shown in Figure A.16, the bode plot of the $\mathrm{HD}_{3,\mathrm{CL}}(j\omega)$ curve will show a maximum somewhere between the open-loop (ω_{p1}) and the closed-loop cut-off frequency ($\omega_{p1}a_1H$). Interestingly enough, the level of maximum distortion does not depend on the feedback factor H, but only on

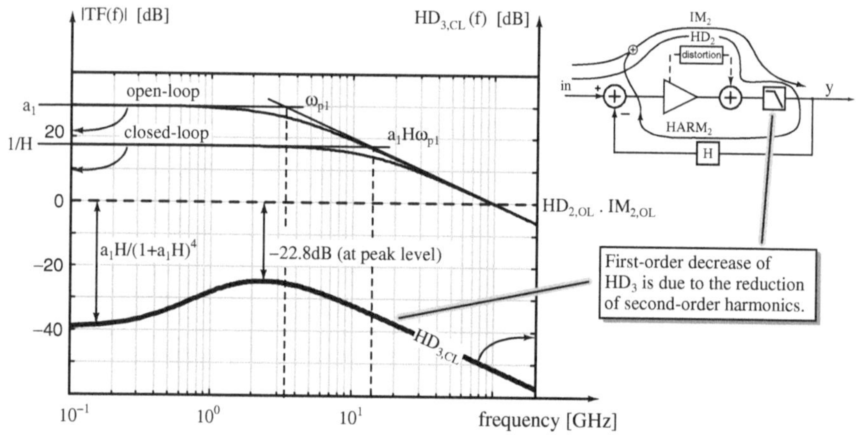

Figure A.16. Frequency dependent third-order harmonic distortion, induced by second-order intermodulation beat products between the input signal and second-order distortion fed back to the input of the system. Note that the third-order distortion caused by the active element itself is ignored in this graph.

the open-loop low-frequency $\mathrm{HD}_{2,\mathrm{OL}}$ and $\mathrm{IM}_{2,\mathrm{OL}}$ characteristics of the amplifier (A.58):

$$\mathrm{HD}_{3,\ \mathrm{CL},\ \mathrm{PEAK}}[\mathrm{dB}] = \mathrm{HD}_{2,\mathrm{OL}}[\mathrm{dB}] + \mathrm{IM}_{2,\mathrm{OL}}[\mathrm{dB}] - 22.8\mathrm{dB} \qquad (A.58)$$

Fortunately, due to the product between $\mathrm{HD}_{2,\mathrm{OL}}$ and $\mathrm{IM}_{2,\mathrm{OL}}$, even the peak magnitude of the feedback-induced third-order distortion remains relatively limited compared to the second-order distortion level. For a wideband amplifier, this is a good thing, since the aggregate amount of both second- and third-order distortion will be reduced. That feedback is not always a good idea, is shown by the following example: consider the case of a narrowband CMOS RF-stage in a receiver front-end. The amplifier has considerable second-order distortion, but at the same time a decent third-order linearity. If not carefully designed, employing local feedback in the amplifier could bring third-order intermodulation into existence, which generates *in-band* distortion. In contrast to this, second-order harmonics can be dealt with in the same way as out-of-band blocker signals, and are effectively removed by the band-select filter in front of the receiver.

Frequency dependent distortion in a MOS amplifier

This section describes the practical limitations of single-stage and multi-stage transistor amplifiers in terms of distortion performance versus maximum usable frequency. It will become clear that in a real-world implementation, there is a trade-off between power, gain, speed and linearity. It turns out that any of those four performance parameters can be improved, but this is always at the expense of increased power consumption or some other compromise in terms of gain, speed or linearity.

Common source transistor amplifier

The first, simple circuit that is examined, is the single-stage common source amplifier. The bias current is provided by a current source with output resistance r_{bias}, while the useful small-signal output current is redirected into a load capacitor C_{load}. Only the nonlinearities of the voltage-to-current conversion characteristic of the transistor will be taken into account at this point. The variation of the output current on the drain voltage of the transistor is ignored for now. It is for this reason that the load capacitor is supposed to be large enough so that all of the small signal AC output current is sinked by the load, while there is only a minimal voltage swing at the drain node of the NMOS transistor (Figure A.17). The second-order distortion components of this circuit can be immediately reduced to the second-order voltage-current characteristic $(V_{\mathrm{GS}} - V_T)^2$ of the transistor.

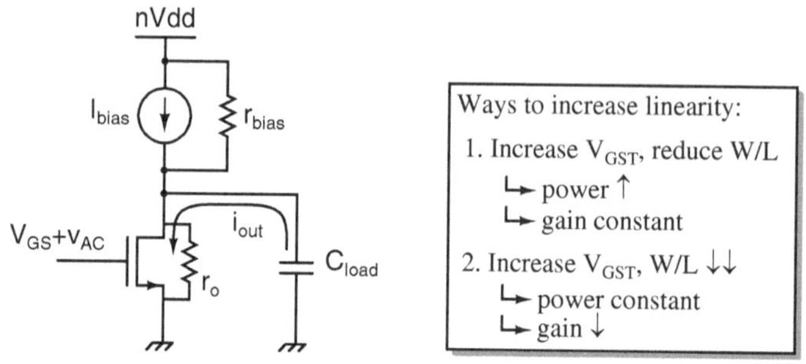

Figure A.17. In the basic CMOS common source amplifier, there is an explicit trade-off between linearity, gain and power consumption. Only nonlinearities of the transconductance are taken into account, while the nonlinear output impedance r_o of the transistor (caused by channel length modulation) is ignored.

A distinction is made between the signal of interest and the bias settings of the circuit. In an analog design, a small AC-signal is commonly superimposed on a larger DC signal, which is called the *operating point* of the transistor. As a result, the signal that is applied to the second-order transconductance characteristic of the transistor can be written as a combination of a fixed DC-setting and a varying AC-part: $(\text{DC} + \text{AC})^2$. If this characteristic is expressed using the Taylor series equivalent, it becomes clear that the magnitude of the second-order component is in fact independent from the operating point of the transistor. The first-order argument, which represents the effective small-signal gain of the transistor, is linearly proportional to the DC-bias value. As a consequence, the linearity performance of a single common-source transistor stage can always be improved by increasing the gate-source overdrive voltage (see Section A.1).

Depending on design-time decisions, this method of improving linearity has inevitable consequences on the performance of the amplifier. For example (Figure A.17), suppose that the overdrive voltage was increased without altering the physical dimensions of the transistor. An increase of the DC-overdrive voltage results in a quadratic increase in drain current Equation (A.2) and thus also in power consumption. The designer can also decide to keep the small-signal gain (g_m) constant. This is achieved by an appropriate reduction of the transistor width. The result is a constant g_m and only a linear increase of the DC bias current. A final choice for the designer is to maintain a constant power dissipation. The consequence however will be a decrease of the transconductance g_m Equation (A.3). Either way, this simple example shows that linearity does not come for free and there is a well-defined trade-off between power, gain and linearity performance.

Resistive degeneration and distortion

The second simple transistor circuit of which the distortion will be examined is the resistively degenerated transistor. The most important properties of this setup were already thoroughly explained in Section A.1, but the dependence on frequency was neglected thus far. If the designer decides to abandon the pure common-source amplifier and introduces some resistance in the source path of the transistor, the current through this resistor causes a signal-dependent voltage drop over the transistor. This results in some kind of local feedback, because the effective voltage which is available over the gate-source terminal of the transistor is decreased. Apart from a reduced overall transconductance of the circuit, the result should be a better linearity performance.

At least in theory, though. While introducing this series resistance, the designer may decide to keep the operating point of the transistor (both the bias current and the overdrive voltage) constant. The DC voltage on all terminals of the transistor will shift proportional to the DC-voltage drop over the degeneration resistor. Apart from the reduction in the small-signal gain Equation (A.28), there are some consequences that arise from this approach: the voltage drop over the resistor can consume a considerable amount of the voltage budget in a deep-submicron design. This may result in additional side effects which could successfully nullify all efforts concerning linearity. For example (Figure A.18), the linearity of the output resistance of a MOS transistor increases when the transistor is biased further into the saturation region (i.e. a larger V_{DS}). A reduction

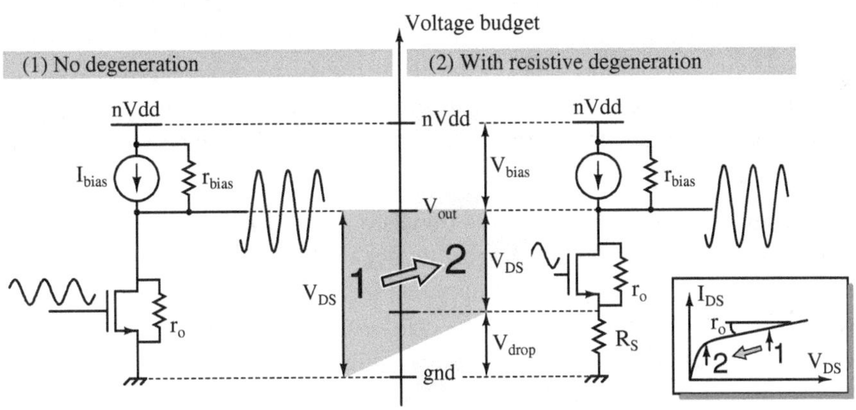

Figure A.18. The introduction of a degeneration resistor in the source path of the CMOS causes a voltage drop. For low voltage technologies, this forces the operation point of the transistor from the saturation region (1) further towards the linear region (2). The nonlinear behaviour of r_o in the new operating point may counteract the benefits of degeneration.

of the voltage headroom would force the designer to bias closer to the linear region. This results in a higher distortion due to channel length modulation in the parasitic output resistance of the transistor due to the AC voltage swing at the output. Note that this effect is not visible if the output current of the transistor immediately forced in the low-impedance of a large capacitive load or in the source terminal of a cascode transistor.

A second possibility for the designer is to maintain a constant power consumption, while the width of the transistor is increased. The bias current is subsequently kept constant by a reduction of the overdrive voltage ($V_{GS} - V_T$). By doing so, the transistor settings shift closer to the weak inversion region as a result of which the small-signal gain (g_m) starts to increase. It should be noted that the improvement of transconductance gain of the transistor is not necessarily translated in a higher voltage gain: a high amount of feedback provided by the degeneration resistor (which is defined by the $g_m R_S$ product) only increases the loop gain of the local feedback mechanism. As a result, the linearity of the resistively degenerated transistor will increase.

The reader should keep in mind that these findings are completely opposite to conclusions that were made for the common-source amplifier stage: remember that in the latter case (without degeneration), it was shown that the overdrive voltage should be increased in order to obtain better linearity (Section A.1). At first sight, it may seem somewhat strange to the reader that without additional power consumption, the linearity performance can be increased merely by enlarging the transistor dimensions. This concern is indeed justified since for a constant bias current, the quadratic transistor model indicates that the transistor dimensions will rise faster than the advantage in terms of transconductance gain. Eventually, the frequency performance will collapse under the heavy weight of parasitics due to the large transistor dimensions. More precisely, the cut-off frequency (f_T) of the transistor, defined by the g_m / C_{GS} ratio, drops to a level much lower than the maximum performance specified for a particular technology.

Apart from a deteriorating f_T of the transistor itself, the increased gate-source capacitance will have a much larger impact on the bandwidth when the amplifier is embedded in a practical application: in combination with the output resistance of the previous stage which is driving the amplifier, the large capacitive load between gate and source creates a pole in the gain characteristic of the input signal source. This finding suggests that, for a fixed power consumption, there is an inevitable trade-off boundary between bandwidth and linearity. The previously described method to improve the linearity of a single transistor amplifier is only useful for low to moderate operating frequencies. Even worse, since a single-stage amplifier can only provide a very limited transconductance

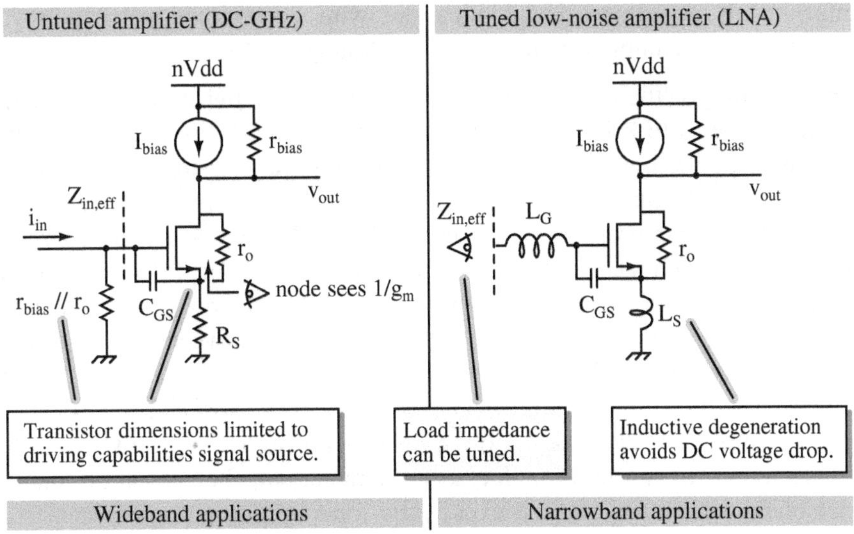

Figure A.19. In a narrowband application, for example an LNA, the parasitics of the transistor can be tuned to the operating frequency by embedding the device in a matching network. Also, inductive degeneration can be used, which avoids a DC voltage drop. In contrast, the signal source of a wideband amplifier directly drives the capacitive load of the transistor.

gain, its possibilities to trade gain for linearity are also limited, which makes this type of amplifier unsuitable for use in low-distortion applications (even in the low-frequency range).

One possible exception to this universal fact is the low-noise amplifier (LNA) in the input stage of a receiver front-end (Figure A.19). The combination of two important conditions makes degeneration a common practice in this type of amplifiers. First of all, the power level that is available from the antenna is much weaker than all other signal found in the remaining part of the receiver. Because of the fairly limited voltage levels that the LNA has to cope with, distortion at the output will automatically stay at acceptable levels.

Secondly, in most cases it turns out that it is possible to tune the amplifier to the frequency of interest. The parasitic capacitance caused by large transistor dimensions can be taken into account during the design of the resonating LC-tank surrounding the transistor. This method is only applicable when the bandwidth of the input signal is low, compared to the center frequency (high fractional bandwidth or Q-factor). This technique can be used up to considerable frequencies, only limited by physical limitations such as the quality factor of the parasitic capacitor or the self-resonance frequency of certain circuit elements.

Unfortunately, it follows that applications with a low fractional bandwidth, such as baseband amplifiers, fall out of this category of lucky few.

A final resort for the analog designer to increase the linearity of a degenerated transistor is to bring extra power into the game. Within certain boundaries, more power implies more transconductance gain for the same transistor dimensions. Unfortunately, the benefits in terms of the current gain cut-off frequency (f_T) are not proportional to the increase in bias current pushed through the transistor. The reason behind this can be easily explained: for a first-order increase of the overdrive voltage, the corresponding increase of the standby current will be found to be somewhere in the region between a linear and an exponential behaviour. The exact figure depends on the actual biasing point of the transistor and deep-submicron oddities such as short channel effects or velocity saturation. The gain of the transistor however, only grows with the first derivative of the transconductance characteristic. For any transconductance characteristic beyond the exponential operating point of the transistor, this implies that the increase in small-signal gain (g_m) is slower than the increase in current consumption.

It is a common pitfall to suppose that the increase in transconductance gain (by increasing the bias current) can be fully devoted to obtain a better linearity performance: a more than linear increase of the DC-bias current will most likely force the designer to reduce the value of the degeneration resistance with the same factor. The result: the steep decrease of the series resistance –necessary to keep the voltage headroom within reasonable boundaries – is not counteracted by the increase of the transconductance gain. The role of the local feedback loop on the linearity performance is thus reduced, with a factor which is larger than could be gained in the first place by increasing the transconductance gain! How come that in spite of these findings, the overall linearity performance is still improved? The correct answer lies in the fact that the overdrive voltage had been increased in order to obtain a larger DC bias current through the transistor. At the same time, the increase of the overdrive voltage results in a decrease of the relative signal swing at the gate terminal of the transistor. As was already pointed out in Section A.1, additional suppression of distortion can be achieved by this measure. The considerable amount of losses both in power and voltage headroom limit the applicability of the resistive degeneration technique in low-distortion high-speed applications.

Remark that, in some cases, the voltage drop over the degeneration resistor can be avoided by replacing the resistor by an reactive element. In case of inductive degeneration, the resistive losses at DC frequencies are avoided, while the desired degenerative properties only come into play at higher frequencies. But once again, this option is only available for narrowband amplifiers, so it still provides no adequate solution for high-performance baseband amplifiers.

Frequency dependent linearity of differential stages

In the previous sections, our focus had been limited to the case of a single-stage single transistor amplifier. However, several multi-transistor configurations have been brought into existence, with a performance exceeding this of a single transistor stage by many orders of magnitude. The first thing that comes to the mind is the differential amplifier. The main advantage of this type of amplifier is that even-order distortion components are suppressed. Most undesired even-order signal components are generated by power-supply noise or external noise sources. This noise signal couples into high-impedance nodes of the circuit, which includes basically every node in the circuit where a considerable voltage gain must be realized. Also, even-order components caused by the transistor pair itself are suppressed, which makes the differential topology suited to counteract distortion caused by the quadratic characteristic which is so characteristic of MOS transistors. Provided that the amplifier is designed with the appropriate symmetry and matching rules in mind, both signal paths of the differential amplifier are affected more or less to the same extent by the noise source. Because only the differential output signal is of interest in a differential configuration, common mode noise can be safely ignored under most circumstances.

In the above discussion, it was described how a symmetrical circuit topology provides immunity against distortion merely by agreeing on the fact that common-mode and differential signals have orthogonal dimensions. An important, but easily underestimated player in a differential topology is the output impedance of the current source, located at the common terminal node of the differential gain pair (Figure A.20). Because this biasing transistor does not play any significant role in the differential current flow between the two gain elements, it does also not appear in the differential small-signal gain expression of the amplifier. Conversely, the bias transistor *does* actually interfere with the common mode current flow, in the sense that it prevents changes in the total net current through the differential pair. The result is that the common-mode transconductance gain is being suppressed. The efficiency of this common-mode suppression heavily depends on the output impedance of the current source.

In the lower frequency range, the (finite) output resistance of the current source plays a leading role in the common mode suppression ratio performance of the differential amplifier. In order to increase the output resistance of the bias transistor, one may choose to extend the channel length. However, increasing the dimensions of the current source must be done with caution: a larger area of the current source transistor will always be accompanied with larger parasitic capacitances. It follows that for higher frequencies, the role of the resistive output impedance is gradually replaced by capacitive parasitics. As a

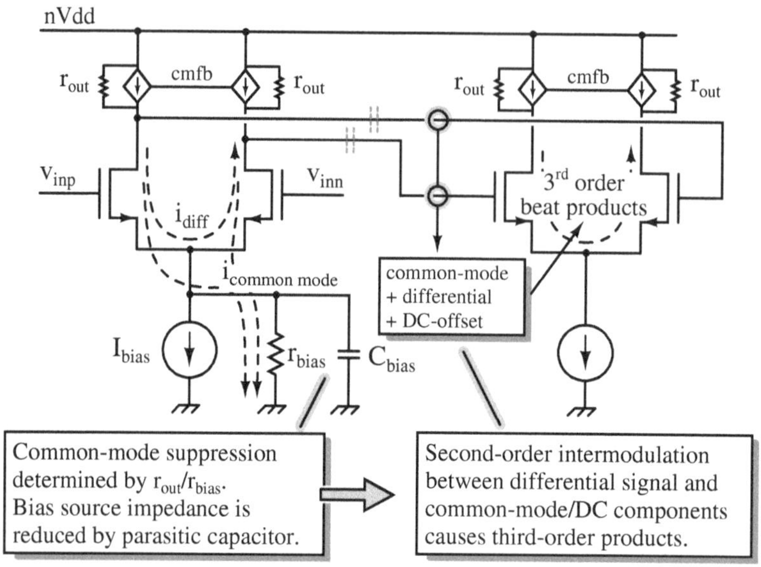

Figure A.20. The common-mode distortion suppression quality in a differential amplifier relies on the impedance of the current source. At high frequencies, the parasitic capacitance causes increasing common mode output components, which are converted to the differential domain by beat products in the next stage.

consequence, the common-mode suppression performance deteriorates for increasing frequencies. At first sight, it may seem to the reader that the decrease of the common-mode rejection capabilities can be safely ignored for as long as the common-mode distortion remains strictly separated from the differential domain.

The real problem in fact, emerges in the subsequent gain stage. The increased common-mode noise power can cause unexpected third-order intermodulation products in the nonlinear transistors of the subsequent amplifier. In other words, even-order distortion, noise or power supply ripple begins to leak into the differential domain. It is emphasized that the *mix products* between both domains have to interpreted in a very wide sense of the word. Also random (but relatively stable) mismatch errors in, for example, the threshold voltage or a difference in mobility between the transistors of a differential amplifier cause small differences in the DC biasing. The mixing products originating from this DC operating point offset and a common-mode interferer will end up in differential intermodulation byproducts corrupting the output spectrum. The latter case, where only the DC offset is taken into account, is also known as the so-called *common-mode to differential conversion gain*. The ratio of the differential gain to the common-mode conversion gain of the amplifier is commonly referred to as the *common mode rejection ratio* (CMRR) of the amplifier. It is a measure for the sensitivity of the amplifier to externally injected interferers or circuit non-idealities such as even-order distortion caused by the transistors themselves.

As pointed out earlier, the parasitic capacitance of the current source introduces a pole in the CMRR characteristic, beyond which the even-order distortion immunity of the differential amplifier is reduced. It must be noted however that the very basic structure of the differential amplifier makes it to one of the fastest topologies to suppress common-mode signals, because the number of transistors included in the signal path is limited to a number of only two transistors.

Distortion performance of multistage amplifiers

This section describes the linearity performance limitations of multistage amplifiers. In contrast to the previously discussed single-stage amplifier, each stage in the multistage amplifier basically includes a transconductance section followed by a current-to-voltage conversion step. The output voltage of one stage is subsequently applied to the input of the subsequent stage. The importance of the I-to-V conversion step should not be underestimated. Remember that in the example of a transconductance amplifier, a large load capacitance prevented a voltage swing at the output of the amplifier (Figure A.17). As a consequence, the nonlinear output resistance of the transistor did not play any significant role in terms of linearity. The output resistance of the active element however plays a leading role in the current-to-voltage conversion process of the multistage amplifier. Unfortunately, the value of the output resistance r_o is susceptible to changes of the drain-source voltage V_{DS} over the transistor, which is – as it happens – inherent to the I-to-V conversion process. The underlying mechanism of this non-ideal output resistance is known as *channel length modulation* [Red65, Sim92].

With CMOS technologies continuing to advance further into the deep submicron range, extremely short channel lengths cause physical side-effects to become increasingly important. As the width of the depletion region at the drain and source implants starts to take up a significant portion of the channel of the MOS transistor, a transistor model merely based on an ideal current source and a parallel resistor is simply not sufficient any more. It is important to understand that resistive degeneration is not capable to resolve distortion issues caused by the nonlinear character of the output resistance: degeneration can only correct the linearity of the transconductance gain. While it affects the effective input voltage applied over the gate oxide capacitor, the output resistance is not included in the local feedback loop of the degeneration resistor. A possible solution could consist in maintaining a constant drain voltage, for example by introducing a cascode transistor on top of the gain transconductance. However, this merely relocates the I-to-V conversion problem to the drain terminal of the cascode transistor. A more convenient solution in this case – and generally for all amplifiers with a decent low-frequency gain – would be to embed the entire multistage structure in a closed-loop feedback system. The purpose of feedback in this case is to push non-idealities of the active gain elements to the background. Once again, the excess gain available in the forward path is thus 'deployed' to take advantage of the better linearity of passive elements in the feedback path of the system.

At this point, the reader must realize that the low-frequency voltage gain of a multistage amplifier is superior to that of a single stage amplifier. It is thus not surprising that this also shines through in the overall accuracy of the amplifier in feedback configuration. At least in the lower frequency region. Since the high gain prerequisite implies that the signal must pass through a chain of several cascaded transistors on it's way to the output, it is pretty interesting to see what kind of impact this has on the speed of the amplifier and – more specifically – on the linearity in higher frequency bands. The following derivation was made under the assumption of a two-stage amplifier, but can be easily expanded to an arbitrary number of gain stages in the forward path of the feedback system. In a typical case, only one of the stages of a two-stage amplifier is dedicated to voltage amplification (Figure A.21). The other one acts as a buffer stage driving the output, which is commonly a capacitive load in CMOS

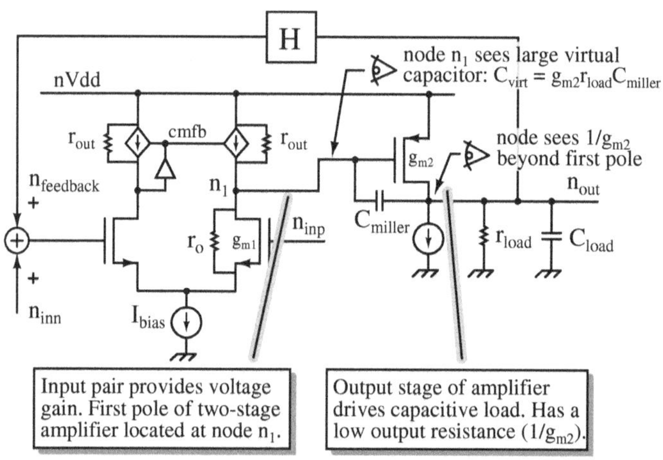

Figure A.21. Example of a two-stage amplifier (CMOS Miller OTA [Ste90]). The output stage determines the second pole (g_{m2}/C_{load}) of the amplifier. Then, taking stability into account, one can determine the GBW-product ($\text{GBW} = g_{m1}/C_{\text{miller}}$). Distortion suppression only starts at frequencies below the first pole of the closed-loop system ($\omega_{\text{p1, CL}} = \text{GBW} \cdot H$).

design. The reason behind this structure is twofold: in order to achieve high voltage gain, the output resistance of the input stage is typically boosted using cascoded transistors. Alas, the high output resistance is equivalent to limited capacitive load driving capabilities. The buffer stage at the output provides less voltage gain, but has a lower impedance for driving a higher capacitive load.

However, this is only one part of the story. Every single stage in the amplifier introduces an extra pole in the signal path. If this second-order system is going to be embedded in a feedback loop, stability analysis comes into play. Without going too much into detail, a stable feedback system requires a certain positive phase margin at unity gain of the open-loop system (which does include the feedback path!). Depending on the exact requirements, a maximally flat frequency response or a minimal settling time, a typical value of the phase margin is in the range of 60–70°. What is of more importance in the context of this discussion is that stability also implies that the location of the poles cannot be chosen arbitrarily. The first pole of the system is always located at a fairly low frequency and causes the decrease of 20 dB/decade (6 dB/octave) in gain. Which of course, in a closed loop system, will be translated in a decrease of excess loop gain and a corresponding decrease of distortion suppression. In order to guarantee stability, the non-dominant pole of the amplifier must be located well beyond the unity-gain crossover point of the open-loop transfer characteristic. A typical value for the location of the second pole is around two or three times beyond the frequency of the open-loop unity-gain crossover point.

At this moment, it would be interesting to find out what frequency can be achieved for the second pole. And then, by performing the previous reasoning backwards, find out what maximum distortion suppression can be achieved up to a certain frequency. First, it is important to remember that the maximum cut-off frequency of one single transistor is defined by f_T. The cut-off frequency defines the point at which, for a certain technology, the *current gain* of a minimum sized transistor drops to unity. This means that beyond this frequency point, the signal source must deliver more current to the gate-source capacitor than what is eventually available at the drain terminal of the transistor.

CUT-OFF FREQUENCY VERSUS $f_{\max}$

It should be noted that this does not mean that the transistor becomes utterly useless for frequencies beyond f_T. For tuned, passband applications, the transistor can serve up to much higher frequencies. This would be defined by the *maximum frequency of oscillation* ($f_{\max}$), but this goes out of the scope of this discussion [Dro55]. In a baseband application however, using a transistor above it's unity gain frequency is a fairly uncommon operation. Although it would give evidence of some original and remarkable insight of the designer.

Of course, the self-loading of a standalone, isolated MOS transistor does not provide a realistic estimate for the practically achievable cut-off frequency. In any real-world environment, the output load of the transistor is not limited to the input capacitance of the subsequent stage: there is such a thing as capacitive wiring load or parasitic output impedance of the active current-supplying transistors. Even the inevitable load of supporting circuits as those monitoring the operating point should be taken into account. The result is that the best achievable crossover frequency of the transconductance gain is only a pale shadow of the f_T reported for a certain CMOS technology. A typical value of the maximum achievable cut-off frequency of a transistor embedded in a circuit is f_T/N, with $N = 5 \ldots 6$. Ideally, for a 0.13 μm CMOS technology with a unity current gain frequency of $f_T = 100\,\text{GHz}$, the second high-frequency pole of the two-stage amplifier is thus located in the range from 15 to 20 GHz. If stability is also brought into the game, the maximum gain-bandwidth product (GBW) of a two-pole feedback system is $f_T/3N$ which is around 5 GHz. If G[dB] is the required gain, the 3 dB cut-off frequency of the closed loop system is given by $f_{3\text{dB, CL}} = \text{GBW [dB Hz]} - \text{G [dB]}$.[5]

The reader should understand that at this point there is still no excess gain available which could be traded for some distortion suppression. Continuing the example from above, suppose that the closed-loop gain was set to 10 dB. In combination with a GBW of 5 GHz, this would mean that nonlinearities of the amplifier are only suppressed for frequencies below 1.58 GHz. Going even further down in frequency, more and more excess gain will become available in the loop of the feedback system, which can be traded for extra linearity. For example, at a frequency of 500 MHz, the excess gain would be 10 dB. Recalling Section A.2, it can be found that this would be enough for an improvement of 20 dB of the HD_2 Equation (A.45) and 30 dB in case of HD_3 Equation (A.47) distortion performance. Remember that all these numbers were derived under the assumption of a technology with an f_T of 100 GHz! From this discussion, it should be clear that a multistage amplifier in a closed-loop configuration is not a suitable candidate for highly-linear applications operating at speeds in the GHz-range.

To conclude this appendix, it is brought to the attention of the reader that the benefit of the cascoding technique in a feedback amplifier depends largely on the actual application. Only for very low speed applications, the larger gain offered by cascoding can contribute to a better linearity in a closed-loop configuration.

[5] Remark that these calculations are somewhat restrictive, since it is supposed that the two-stage amplifier must be stable in unity feedback conditions.

References

[And00] C. Andren, K. Halford, and M. Webster, "CCK, the new IEEE 802.11 standard for 2.4GHz wireless LANs", in *Proceedings International IC-Taipei Conference*, pp. 25–39, May 2000.

[Ass01] Telecommunications Industry Association, "TIA/EIA-644-A: Electrical characteristics of low voltage differential signalling (LVDS) interface circuits", 2001.

[Bah74] L. Bahl, J. Cocke, F. Jelinek, and J. Raviv, "Optimal decoding of linear codes for minimizing symbol error rate", in *IEEE Transactions on Information Theory*, volume 20(2), pp. 284–287, March 1974.

[Bar05] F.J.B. Barros, R.D. Vieira, and G.L. Siqueira, "Relationship between delay spread and coherence bandwidth for UWB transmission", in *Proceedings of the IEEE International Conference on Microwave and Optoelectronics*, pp. 415–420, July 2005.

[Bei01] F.E. Beichelt and P.L. Fatti, *Stochastic Processes and Their Applications*, CRC Press, 2001, ISBN 0415272327.

[Ben02] N. Benvenuto, G. Cherubini, and U. Cherubini, *Algorithms for Communications Systems and Their Applications*, Wiley, 2002, ISBN 0470843896.

[Ber93] C. Berrou, A. Glavieux, and P. Thitimajshima, "Near shannon limit error-correcting coding and decoding: Turbo-codes. 1", in *IEEE International Conference on Communications*, volume 2, pp. 1064–1070, May 1993.

[Bur98] A.G. Burr, "Wide-band channel modelling using a spatial model", in *Proceedings of Spread Spectrum Techniques and Applications*, volume 1, pp. 255–257, September 1998.

[Cai98] G. Caire, G. Taricco, and E. Biglieri, "Bit-interleaved coded modulation", in *IEEE Transactions on Information Theory*, volume 44(3), pp. 927–946, May 1998.

[CEN97] *CENELEC prEN 50067:1997, Specification of the radio data system (RDS)*, published jointly by CENELEC (rue de Stassart, B-1050 Brussels, Belgium) and EBU, Geneva, update of the 1992 edition, 1997.

[Cha79] U. Charash, "Reception through nakagami fading multipath channels with random delays", in *IEEE Transactions on Communications*, volume 27(4), pp. 657–670, April 1979.

[Cha00] A. Chandrakasan, W.J. Bowhill, and F. Fox, *Design of High-Performance Microprocessor Circuits*, Wiley-/IEEE Press, 2000, ISBN 078036001X.

[Com96] D.J. Comer, "A theoretical design basis for minimizing CMOS fixed taper buffer area", in *IEEE Journal of Solid-State Circuits*, volume 31(6), pp. 865–868, June 1996.

[Com02] Federal Communications Commission, "Revision of part 15 of the commission's rule regarding ultra-wideband transmission systems, first report and order, ET Docket 98-153, FCC 02-48", 2002.

[Con00] J. Conover, "Anatomy of IEEE 802.11b wireless", *[Online document] URL: http://www.networkcomputing.com/1115/1115ws2.html*, August 2000.

[Coo93] J.W. Cooley, and J.W. Tukey, "On the origin and publication of the FFT paper", in *Current Contents*, volume 33(51–52), pp. 8–9, December 1993.

[Cou97] L.W. Couch, *Digital and Analog Communication Systems*, Prentice Hall, fifth edition, 1997, ISBN 0-13-522583-3.

[Cox72] D.C. Cox, "Delay doppler characteristics of multipath delay spread and average excess delay for 910 MHz urban mobile radio paths", in *IEEE Transactions on Antennas and Propagation*, volume AP-20(5), pp. 625–635, September 1972.

[Cox75] D.C. Cox and R.P. Leck, "Distributions of multipath delay spread and average excess delay for 910 MHz urban mobile radio paths", in *IEEE Transactions on Antennas and Propagation*, volume AP-23(5), pp. 206–213, March 1975.

[Dav71] B. Davis, "FM noise with fading channels and diversity", in *IEEE Transactions on Communications*, volume 19(6), pp. 1189–1200, December 1971.

[Ded07] J. Dederer, B. Schleicher, F. De Andrade Tabarani Santos, A. Trasser, and H. Schumacher, "FCC compliant 3.1–10.6 GHz UWB pulse radar system using correlation detection", in *IEEE/MTT-S International Microwave Symposium*, pp. 1471–1474, June 2007.

[Des03] C. Desset and A. Fort, "Selection of channel coding for low-power wireless systems", in *IEEE Transactions on Vehicular Technology*, volume 3, pp. 1920–1924, April 2003.

[DeV61] A.J. DeVries, "Design of stereophonic receiver for a stereo system in the FM band using an AM subcarrier", *Transactions on IRE*, volume BTR-7(2), pp. 67–72, July 1961.

[Dev87] D. Devasirvatham, "Multipath time delay spread in the digital portable radio environment", in *IEEE Communications Magazine*, volume 25(6), pp. 13–21, June 1987.

[Dev90] D. Devasirvatham, M.J. Krain, D.A. Rappaport, and C. Banerjee, "Radio propagation measurements at 850 MHz, 1.7 GHz and 4 GHz inside two dissimilar office buildings", in *Electronics Letters*, volume 26(7), pp. 445–447, March 1990.

[Dro55] P. Drouilhet, "Predictions based on maximum oscillator frequency", in *IRE Transactions on Circuit Theory*, volume 2(2), pp. 178–183, June 1955.

[Duo06] Q. Duong, Q. Le, C. Kim, and S. Lee, "A 95-db linear low-power variable gain amplifier", in *IEEE Transactions on Circuits and Systems I: Regular Papers*, volume 53(8), pp. 1648–1657, August 2006.

[Eng02] M. Engels, *Wireless OFDM Systems: How to Make Them Work?* Kluwer, 2002, ISBN 1-4020-7116-7.

[Eth05] "802.3-2005: IEEE standard for information technology – telecommunications and information exchange between systems – local and metropolitan area networks – specific requirements – part 3: Carrier sense multiple access with collision detection (CSMA/CD) access method and physical layer specifications", 2005, Revision of IEEE Std 802.3-2002.

[Eur90] European Telecommunications Standards Institute, *ETSI TS 100 961: GSM 06.10 Full Rate Speech Transcoding*, January 1990.

[Eur01] European Telecommunications Standards Institute, *ETSI TS 101 475: Broadband Radio Access Networks; HIPERLAN/2; Physical layer*, 2001.

[Eur05] European Telecommunications Standards Institute, *ETSI TS 100 910: GSM05.05 Digital cellular telecommunications system (Phase 2+); Radio Transmission and Reception*, November 2005.

[Eyn89] F. Op't Eynde and W. Sansen, "Design and optimisation of CMOS wideband amplifiers", in *Proceedings of the IEEE Custom Integrated Circuits Conference*, pp. 25.7/1–25.7/4, May 1989.

[Fel45] W. Feller, "The fundamental limit theorems in probability", in *Bulletin of the American Mathematical Society*, volume 51(11), pp. 800–832, 1945.

[Foc01] G. Fock, J. Baltersee, P. Schulz-Rittich, and H. Meyr, "Channel tracking for RAKE receivers in closely spaced multipath environments", in *IEEE Journal on Selected Areas in Communications*, volume 19(12), pp. 2420–2431, December 2001.

[Fon04] R.J. Fontana, "Recent system applications of short-pulse ultra-wideband (UWB) technology", in *IEEE Transactions on Microwave Theory and Techniques*, volume 52(9), pp. 2087–2104, September 2004.

[For96] G.D. Forney, L. Brown, M.V. Eyuboglu, and J.L. Moran, "The V.34 high speed modem standard", in *IEEE Communications Magazine*, volume 34(12), pp. 28–33, December 1996.

[Fre99] S. Freisleben, A. Bergmann, U. Bauernschmitt, C. Ruppel, and J. Franz, "A highly miniaturized recursive Z-path SAW filter", in *Proceedings Ultrasonics Symposium*, pp. 347–350, 1999.

[Fri46] H.T. Friis, "A note on a simple transmission formula", in *Proceedings of the IRE*, volume 34(5), pp. 254–256, May 1946.

[Fuj01] K. Fujimoto and J.R. James, *Mobile Antenna Systems Handbook*, Artech House, second edition, 2001, ISBN 1580530079.

[Gas05] M.S. Gast, *802.11 Wireless Networks: The Definitive Guide*, O'Reilly Media, Inc., 1005 Gravenstein Highway North, Sebastopol, CA 95472, second edition, 2005.

[Gav84a] J.J. Gavan and M.B. Shulman, "Effects of densensitization on mobile radio system performance, part I: Qualitative analysis", in *IEEE Transactions on Vehicular Technology*, volume 33(4), pp. 285–290, November 1984.

[Gav84b] J.J. Gavan and M.B. Shulman, "Effects of densensitization on mobile radio system performance, part II: Quantitative analysis", in *IEEE Transactions on Vehicular Technology*, volume 33(4), pp. 291–300, November 1984.

[Gil68] B. Gilbert, "A precise four-quadrant multiplier with subnanosecond response", in *IEEE Journal of Solid-State Circuits*, volume 3(4), pp. 365–373, December 1968.

[Han67] F. Hansen, "Desensitization in transistorized PM/FM-receivers", in *IEEE 18th Vehicular Technology Conference*, volume 18, pp. 78–86, December 1967.

[Han03] L. Hanzo, B.J. Choi, T. Keller, and M. Münster, *OFDM and MC-CDMA for Broadband Multi-user Communications, WLANs and Broadcasting*, Wiley, 2003, ISBN 0470858796.

[Har69] P. Hartemann and E. Dieulesaint, "Acoustic-surface-wave filters", in *Electronics Letters*, volume 5, pp. 657–658, December 1969.

[Has97] R. Hassun, M. Flaherty, R. Matreci, and M. Taylor, "Effective evaluation of link quality using error vector magnitude techniques", in *Proceedings of Wireless Communications Conference*, pp. 89–94, August 1997.

[Hee01] C. Heegard, J.T. Coffey, S. Gummadi, P.A. Murphy, R. Provencio, E.J. Rossin, S. Schrum, and M.B. Shoemake, "High performance wireless ethernet", in *IEEE Communications Magazine*, volume 39(11), pp. 64–73, November 2001.

[Hei73] G.L. Heither, "Characterization of nonlinearities in microwave devices and systems", in *IEEE Transactions on Microwave Theory and Techniques*, pp. 797–805, December 1973.

[Hol99] C.L. Holloway, M.G. Cotton, and P. McKenna, "A model for predicting the power delay profile characteristics inside a room", in *IEEE Transactions on Vehicular Technology*, volume 48(4), pp. 1110–1120, July 1999.

[IT93] ITU-T, "Recommendation V.32 for a family of 2-wire, duplex modems operating at data signalling rates of up to 9600 bit/s for use on the general switched telephone network and on leased telephone-type circuits", March 1993.

[Jak94] W.C. Jakes, *Microwave Mobile Communications*, IEEE Press Classic Reissue Series, Wiley, 1994, ISBN 0780310691.

[Kam97] A. Kamerman and N. Erkocevic, "Microwave oven interference on wireless LANs operating in the 2.4 GHz ISM band", in *The 8th IEEE International Symposium on Personal, Indoor and Mobile Radio Communications (PIMRC '97)*, pp. 1221–1227, September 1997.

[Kim08] J.S. Kim, S.C. Kim, S.S. Hwang, B. Kang, and J.-S. Park, "Performance enhancement of timing acquisition and detection in UWB-IR matched filter receiver", in

ICACT 2008. 10th International Conference on Advanced Communication Technology, volume 2, pp. 1347–1351, February 2008.

[Lak94] K.R. Laker and W. Sansen, *Design of Analog Integrated Circuits and Systems*, McGraw-Hill, 1994, ISBN 0-07-113458-1.

[Lee04] S. Lee, S. Bagga, and W.A. Serdijn, "A quadrature downconversion autocorrelation receiver architecture for UWB", in *International Workshop on Ultrawideband Systems*, pp. 6–10, May 2004.

[Lee06] H.D. Lee, K.A. Lee, and S. Hong, "Wideband VGAs using a CMOS transconductor in triode region", in *36th European Microwave Conference*, pp. 1449–1452, September 2006.

[Lin04] S. Lin and D.J. Costello, *Error Control Coding*, Pearson Education/Prentice Hall, second edition, 2004, ISBN 013-017973-6.

[Lun74] W.N. Lundberg and C. von Lanzenauer, "The n-fold convoluation of a mixed density and mass function", in *ASTIN Bulletin International Actuarial Association – Brussels, Belgium*, volume 8(1), pp. 91–103, 1974.

[Mag01] G.M. Maggio, N. Rulkov, and L. Reggiani, "Pseudo-chaotic time hopping for UWB impulse radio", in *IEEE Transactions on Circuits and Systems I: Fundamental Theory and Applications*, volume 48(12), pp. 1424–1435, December 2001.

[Mat04] P. Mattheijssen, M.H.A.J. Herben, G. Dolmans, and L. Leyten, "Antenna-pattern diversity versus space diversity for use at handhelds", in *IEEE Transactions on Vehicular Technology*, volume 53(4), pp. 1035–1042, July 2004.

[McE84] R. McEliece and W. Stark, "Channels with block interference", in *IEEE Transactions on Information Theory*, volume 30(1), pp. 44–53, January 1984.

[Muh05] K. Muhammad, Y.C. Ho, T. Mayhugh, C.M. Hung, T. Jung, I. Elahi, C. Lin, I. Deng, C. Fernando, J. Wallberg, S. Vemulapalli, S. Larson, T. Murphy, D. Leipold, P. Cruise, J. Jaehnig, M.C. Lee, R.B. Staszewski, R. Staszewski, and K. Maggio, "A discrete time quad-band GSM/GPRS receiver in a 90nm digital CMOS process", in *Proceedings of the IEEE 2005 Custom Integrated Circuits Conference*, pp. 809–812, September 2005.

[O'D05] I.D. O'Donnell and R.W. Brodersen, in *IEEE Transactions on Vehicular Technology*, volume 54(5), pp. 1623–1631, September 2005.

[Opp03] A.V. Oppenheim, R.W. Schafer, and J.R. Buck, *Discrete-Time Signal Processing*, Prentice Hall, 2003, ISBN 0-13-754920-2.

[P8007] IEEE P802.15, "P802.15-03/268r3: Multiband OFDM physical layer proposal for IEEE 802.15 task group 3a", March 2007.

[Pas79] S. Pasupathy, "Minimum shift keying: A spectrally efficient modulation", in *IEEE Communications Magazine*, pp. 14–22, July 1979.

[Pel89] M.J.M. Pelgrom, A.C.J. Duinmaijer, and A.P.G. Welbers, "Matching properties of MOS transistors", in *IEEE Journal of Solid-State Circuits*, volume 24(5), pp. 1433–1439, October 1989.

[Pie90] S.S. Pietrobon, R.H. Deng, A. Lafanechere, G. Ungerboeck, and D.J. Costello, "Trellis-coded multidimensional phase modulation", in *IEEE Transactions on Information Theory*, volume 36(1), pp. 63–89, January 1990.

[Pin01] L. Ping, X. Huang, and N. Phamdo, "Zigzag codes and concatenated zigzag codes", in *IEEE Transactions on Information Theory*, volume 47(2), pp. 800–807, February 2001.

[Poz05] D.M. Pozar, *Microwave Engineering*, Wiley, third edition, 2005, ISBN 0-471-17096-8.

[Pro00] J.G. Proakis, *Digital Communications*, McGraw-Hill, fourth edition, 2000, ISBN 0072321113.

[Rap02] T.S. Rappaport, *Wireless Communications: Principles and Practice*, Prentice Hall, second edition, 2002, ISBN 0130422320.

[Raz94] B. Razavi, Y. Ran-Hong, and K.F. Lee, "Impact of distributed gate resistance on the performance of MOS devices", in *IEEE Transactions on Circuits and Systems I: Fundamental Theory and Applications*, pp. 750–754, November 1994.

[Red65] V.G.K. Reddi and C.T. Sah, "Source to drain resistance beyond pinch-off in metal-oxide-semiconductor transistors (MOST)", in *IEEE Transactions on Electron Devices*, volume 12(3), pp. 139–141, March 1965.

[Rod95] D. Rodriguez, A. Rodriguez, and N.G. Santiago, "On the implementation of fast algorithms for linear codes using T805 microcomputer arrays", in *Proceedings of the 38th Midwest Symposium on Circuits and Systems*, volume 2, pp. 1284–1287, August 1995.

[Ros76] I.G. Rosenberg, "Some algebraic and combinatorial aspects of multiple-valued circuits", in *Proceedings of the Sixth International Symposium on Multiple-Valued Logic*, pp. 9–23, May 1976.

[Rud06] R.F. Rudd, "Physical-statistical model for prediction of power delay profile of indoor radio channels", in *Electronics Letters*, volume 42(17), pp. 957–958, August 2006.

[San99] W. Sansen, "Distortion in elementary transistor circuits", in *IEEE Transactions on Circuits and Systems II: Analog and Digital Signal Processing*, volume 46(3), pp. 315–325, March 1999.

[Sar93] R. Sarpeshkar, T. Delbruck, and C.A. Mead, "White noise in MOS transistors and resistors", in *IEEE Circuits and Devices Magazine*, volume 9(6), pp. 23–29, November 1993.

[Set06] V. Sethuraman and B. Hajek, "Comments on 'bit-interleaved coded modulation", in *IEEE Transactions on Information Theory*, volume 52(4), pp. 1795–1797, April 2006.

[Sha48] C.E. Shannon, "A mathematical theory of communication", in *The Bell System Technical Journal*, pp. 379–423, 623–656, July, October 1948.

[Sib02] S. Sibecas, C.A. Corral, S. Emami, and G. Stratis, "On the suitability of 802.11a/RA for high-mobility DSRC", in *IEEE 55th Vehicular Technology Conference*, volume 1, pp. 229–234, 2002.

[Sim92] T.K. Simacek, "Simulation and modeling", in *IEEE Circuits and Devices Magazine*, volume 8(3), pp. 7–8, May 1992.

[Skl01] B. Sklar, *Digital Communications: Fundamentals and Applications*, Prentice Hall, second edition, 2001, ISBN 0130847887.

[Sou94] E.S Sousa, V.M. Jovanovic, and C. Daigneault, "Delay spread measurements for the digital cellular channel in Toronto", in *IEEE Transactions on Vehicular Technology*, volume 43(4), pp. 837–847, November 1994.

[Ste90] M.S.J. Steyaert and W.M.C Sansen, "Power supply rejection ratio in operational transconductance amplifiers", in *IEEE Transactions on Circuits and Systems*, volume 37(9), pp. 1077–1084, September 1990.

[Swe05] E.G. Swedin and D.L. Ferro, *Computers: The Life Story of a Technology*, Greenwood, Westport, CT 06881, 2005.

[Tag07] D. Taggart, R. Kumar, Y. Krikorian, G. Goo, J. Chen, R. Martinez, T. Tam, and E. Serhal, "Analog-to-digital converter loading analysis considerations for satellite communications systems", in *Proceedings IEEE Aerospace Conference*, pp. 1–16, March 2007.

[Tho00] S. Thoen, L. Van der Perre, B. Gyselinckx, M. Engels, and H. De Man, "Predictive adaptive loading for HIPERLAN/II", in *IEEE 52nd Vehicular Technology Conference*, volume 5, pp. 2166–2172, 2000.

[Tse95] C.K. Tse, S.C. Wong, and M.H.L. Chow, "On lossless switched-capacitor power converters", in *IEEE Transactions on Power Electronics*, volume 10(3), pp. 296–291, May 1995.

[Tse05] D. Tse and P. Viswanath, *Fundamentals of Wireless Communication*, Cambridge University Press, 2005.

[Ung82] G. Ungerboeck, "Channel coding with multilevel/phase signals", in *IEEE Transactions on Information Theory*, volume 28(1), pp. 55–67, January 1982.

[Ver06] W. Vereecken and M. Steyaert, "Interference and distortion in pulsed ultra wideband receivers", in *Proceedings of the IEEE International Conference Ultra-Wideband (ICUWB)*, pp. 465–470, September 2006.

[Vit67] A.J. Viterbi, "Error bounds for convolutional codes and an asymptotically optimum decoding algorithm", in *IEEE Transactions on Information Theory 13*, volume 13(2), pp. 260–269, April 1967.

[Wal99] R.H. Walden, "Analog-to-digital converter survey and analysis", in *IEEE Journal on Selected Areas in Communications*, volume 17(4), pp. 539–550, April 1999.

[Wei87] L.-F. Wei, "Trellis-coded modulation with multidimensional constellations", in *IEEE Transactions on Information Theory*, volume 33(4), pp. 483–501, July 1987.

[Wla07] "802.11-2007: IEEE standard for information technology – telecommunications and information exchange between systems – local and metropolitan area networks – specific requirements – part 11: Wireless LAN medium access control (MAC) and physical layer (PHY) specifications", 2007, Revision of IEEE Std 802.11-1999.

[Zim80] H. Zimmermann, "OSI reference model âĂŤ the ISO model of architecture for open systems interconnection", in *IEEE Transactions on Communications*, volume 28(4), pp. 425–432, April 1980.

Index

FSC
www.fsc.org
MIX
Papier | Fördert
gute Waldnutzung
FSC® C083411

Kolibri 360 GmbH
Ferdinand-Jühlke-Straße 7
99095 Erfurt, Deutschland
produktsicherheit@kolibri360.de